To Mikhaie ~ith
best wishes

12/14/10

THE ENERGY IMPERATIVE

Phil Rae & Leonard Kalfayan
with Michael J. Economides

**With
Compliments of**

PUBLISHING

ET Publishing
Houston, TX

Energy Tribune Publishing Inc.
800 Gessner Rd. - Ste. 1220
Houston, TX 77024

Graphic design and production: Seth Myers and Tony Rose
Production manager: Alexander M. Economides
Copy Editor: Stephanie Weiss
Cover Art: Tony Rose

Published by: Energy Tribune Publishing Inc.
               800 Gessner Rd. - Ste. 1220
               Houston, TX 77024

               (713) 647-0903
               (713) 647-0940 *(fax)*

for orders and customer service enquires
contact: info@energytribune.com

Library of Congress Control Number: 2010934268

ISBN 978-0-578-06501-4

Printed and bound by R.R. Donnelley

Praise for **The Energy Imperative**

The Energy Imperative is the ultimate compendium for the world's most important commodity, energy, so critical for the growth and success of all the world's economies. This must-read book provides an in-depth, eye-opening account of energy today and in the future including products, sources, production, power generation, transmission, storage and usage. It properly sheds light on the energy industry profits in comparison to other industries and quantifies the true picture of risks versus rewards for the sources targeted by environmentalists, media and governments alike.

The authors map out a blueprint for a realistic, responsible and sound energy policy for the future. It is in sharp contrast to that of the US Academies of Science and Engineering "America's Energy Future", allegedly a synthesis of expert opinions which is nothing more than a politically driven analysis based on flawed science recommending misguided policies which would provide a disservice to the nation and the world all in the name of social engineering.

> **– Joseph S. D'Aleo, Fellow of the American Meteorological Society**
> **Executive Director ICECAP US**

Washington DC, the media and pressure groups seem determined to impose an agenda fraught with serious and even grave economic and national security implications, without apparent regard for what they must claim in order to win. The target of this campaign is abundant, affordable energy, viewed in many quarters as a horror to be controlled. Abundant energy is best manifested by "fossil fuels", or hydrocarbons, which "The Energy Imperative" illuminates from its cradle to its grave – what is a fossil fuel, how it is obtained and exploited, to what purpose and value for humanity – a basic education which, sadly, remains a mystery to most voters and increasingly to their peril.

Climate change, physical scarcity, the supposed promise of long-suffering "alternative" energy sources and other claims are promiscuously showered onto the public, with misleading rhetoric and images, and the authors tackle many of the higher profile examples of which with authority and their own perspectives. "*The Energy Imperative*" is an important addition to recent tools for those interested in sorting through the hype and noise.

> **– Christopher C. Horner, Esq., senior fellow with the Competititve Enterprise**
> **Institute and author of "Power Grab**

It's the Energy, Stupid!

Former President Clinton said, " it's the economy, stupid" but he could equally have said energy, instead. It's the same thing because energy is the economy. Without energy, nothing moves, nothing works, nothing gets done, there's nothing to sell and no money to buy it with anyway. People who like to know what they're talking about and like even better to know when politicians, newspapers and global warming activists don't, will love this book. It's very good and needed: fact based and full of thought. The world needs a fact based book on energy. It's a must-read for anyone who really cares about the future of our planet and everyone who relies on its most important industry.

> **– Joe Mach, Inventor of NODAL Analysis, Former Head of Yukos E&P and former**
> **Vice President, Schlumberger**

(continued on back)

I rarely read a book that is both lucid and satisfying in the completeness and logic of its message. *The Energy Imperative* by Rae, Kalfayan and Economides, has succeeded in the difficult task of describing the role of energy in the world today. The authors describe in layman's terms, but without compromising in accuracy and veracity, the physics of the issues and dispel the confusion and ideological dogma existing in some parts of the press and current energy policies.

Hype propagated by Peak oil theorists and media predicting we will quickly exhaust our oil reserves are invalidated by physics and facts.

Often confused and mixed together by the media and environmentalists the issues of carbon neutral electrical energy (wind, solar, nuclear) and transportation energy (oil) are separated and discussed rationally.

The authors make a clear case for why natural gas should, and will, increase in importance. Electrical energy as the authors point out is the most versatile end-use energy and in the future will be used even more by the masses;   the resulting energy intensity will bring the world to a more environmentally sound position.

The Energy Imperative has given me a clear understanding of our choices for the future and will be a reference book for some time to come. I strongly recommend this book to anyone who is interested in energy, the environment or world politics.

> **– Don Wolcott is the COO of Aurora Oil LLC, a Moscow based independent oil producer.**

The need for energy to fuel human economic development is self-evident and has been well documented by numerous authors. What challenges human-kind is how we will fill that need going forward. This is the *Energy Imperative* that the authors address with a balanced review of all the competing alternatives. Data presented allow the reader to develop an objective opinion and to consider that this reviewer's industry – the production of clean-burning natural gas – has a much greater role to play in supplying future needs.

> **– Mark Pearson, President, Energy Investment Consulting Services**

TABLE OF CONTENTS

PREFACE III

FOREWORD VII

INTRODUCTION
The Importance of Energy IX

CHAPTERS

1   What Are Fossil Fuels and Where Are They? 1

2   Who Controls the Oil and Gas? 27

3   Energy Use – Reality 47

4   Well Construction and Production 63

5   Bulk Transportation and Facilities 91

6   Oil Refining and Gas Processing 113

7   The Role of Coal 143

8   Power Generation 159

9   Alternative Transportation Fuels 189

10 Pollution and the Environment 217

11 Climate Change 235

12 Energy Future 257

APPENDICES 277

A   A Few Energy Factlets 279

B   Physical Classification of Crude Oils 283

C   Chemical Classification of Hydrocarbons in Crude Oil 285

D   Oil Refinery Units and Processes 288

E   Gas Processing 293

F   Nuclear Power Plants In Operation and Planned 295

REFERENCES 299

INDEX 303

PREFACE

Jacques Bosio, Steve Holditch, Giovanni Paccaloni and Ed Runyan

For almost four decades, energy with its connection to the world economy, its geopolitical dimensions and, as of late, its future, has dominated the international debate.

We understand the energy industry well and all four of us have spent our working lives in it as petroleum engineers. We also achieved some repute, as all of us have served, at different times, as president of the international Society of Petroleum Engineers. While Big Oil has often been the favorite whipping boy of a number of politicians and other social commentators the world over, our view is that oil (and gas and coal), referred to as fossil fuels, have had a profoundly beneficial impact on modern society and they will continue to do so for many decades to come.

We do not downplay the challenges that arise from the use of these fossil fuels. We are acutely aware of the technological and logistical challenges that are associated with their production and use and we are intimately aware of the environmental risks. All of us had to contend with these risks and their remediation throughout our careers.

Nor are we here to cast aspersions against alternative energy sources such as wind, solar, biofuels and the like. But it is irresponsible for people to be suggesting that these energy sources can replace even a small portion of the energy provided by fossil fuels anytime soon, if ever. Some nuggets from the book "Power Hungry" from our friend Robert Bryce (the Managing Editor of the *Energy Tribune*):

- The world is using 200 million barrels per day of oil equivalent, or about 23.5 Saudi Arabia's production per day. Thus, if the world's policy makers really want to quit using carbon-based fuels, then we will need to find the energy equivalent of 23.5 Saudi Arabia's production every day, and all of that energy must be carbon free.

- It takes 10,000 tons of wood to produce one megawatt of electricity and, if this occurred, the US will be chopping down forests faster than it can grow them.

- For a wind farm it will take 45 times the land mass of a nuclear power station to produce a comparable amount of power and this will also need hundreds of miles of high-voltage lines to get the energy to customers. And it is not just the cost, the concrete and the steel. Wind (and solar) are notoriously intermittent. Texas has built 10,000 MW of wind farms and yet, according to the Electric Reliability Council of Texas, which runs the state's electricity grid, just "8.7 percent of the installed wind capability can be counted on as dependable capacity during the peak demand period."

- Biofuels, especially corn-based ethanol are even worse. The energy density of biofuels, measured in Watts per square foot (of planted corn and the like) is actually less than those for wind and solar, with the latter of course suffering from their intermittent nature.

We do get upset and cannot understand why politicians and news commentators present as facts things that simply cannot happen. These include notions that solar will account for, e.g., 20% of the world energy mix in twenty years or that wind can play a comparable role even sooner. Neither of these energy sources can account for much more than a few percent of world energy over the next twenty years.

None of the above comments is meant to suggest that progress will not be made. The world will have to change and we are quite aware of the fact that oil and gas are exhaustible resources. But changes, most likely favoring variants of nuclear developments and massive infrastructural re-adjustments, such as electrifying private vehicles on a scale that is currently unfathomable, will take many decades to happen. Energy transitions are painful and slow and the economy, not government fiat, will dictate their coming. Of course, we strongly endorse research and development in all types of energy and, especially, energy efficiency. This is both prudent and, potentially, profitable.

In the meantime oil and gas are here to stay.

Oil

While estimates vary, approximately 6 to 8 trillion barrels of oil from conventional resources, with the same or more from unconventional (tar sands, shale oil et al) resources are believed to be 'in-place' and, as yet, untapped. Since the first wells were drilled in the mid-1800s, the world has consumed just one trillion barrels of oil from conventional reservoirs out of this total. Given around the same amount of so-called unconventional oil -- 6 to 8 trillion barrels is estimated to be 'in-place' -- that offers a grand total of around 12 to 16 trillion barrels of oil as yet untapped. Not all in-place oil is, of course, recoverable but continued technology developments will surely increase the percentage that is.

Wolfgang Schollenberger, retired technology VP for BP, writing in Oil Gas European Magazine in January 2006 said that while global hydrocarbon reserves were 4.5 trillion barrels of oil equivalent, after as much as 6 trillion barrels equivalent of oil and gas are produced in the twenty first century, "there will be enough hydrocarbons left so that somebody might estimate on Jan. 1, 2101 the then remaining reserves plus resources to be ... 4.7 trillion barrels of oil equivalent ..."

In 2009, world oil consumption hit 31 billion barrels per year, or around 85 million barrels a day, figures that will continue to rise in the next few years. However, new technology is already having a key impact on oil recovery rates. And while "peak oil" alarmists regularly claim the oil reserves of the current key players, especially Saudi Arabia, are over-estimates, it is especially in this arena that we are already seeing the impact of new technology.

Natural Gas

In November 2009 the International Energy Agency (IEA) in Paris released its world outlook report. In the report it said: "The long-term global recoverable gas resource base is estimated at more than 850 trillion cubic meters."That translates to just over 30,000 trillion cubic feet of gas, which means that in just one year the IEA doubled its estimate of 400 trillion cubic meters from its 2008 report.

The enormous difference in the IEA estimates is almost solely because of the addition of what, for decades, was labeled as unconventional gas, and the new jewel in the crown is the success of shale gas, arguably the shiniest recent success in the petroleum industry.

There is a lot of gas in the world.

At current world consumption of about 105 trillion cubic feet per year, the proven reserves can provide for over 60 years. This to some is comforting and to others, remembering the IEA's number, it is even more comforting, implying a supply that could last almost 300 years. If one only just fantasizes any future contributions from the orders-of-magnitude larger resource in the form of natural gas hydrates, it is easy to see how natural gas is almost certain to evolve into the premier fuel of the world economy.

This book takes the attitude that oil and gas will continue to dominate the world energy scene for decades, spanning the entire twenty first century. The book describes these resources at length and provides the state of the art in their exploitation, mindful of the roles of evolving technology and the need for environmental vigilance.

Because of its long-lasting future, the petroleum industry must be strongly committed to avoid serious damage to the environment and to establish a pervasive culture of health, safety and environmental (HSE) conduct. The best energy companies of the future will be appraised strictly on their "technological ability" of exploring and producing hydrocarbons while guaranteeing extremely high standards of HSE. There will be very little, to hopefully zero, tolerance for major environmental disasters to take place. We, as a group of past SPE presidents, do solicit discussion on how to ensure that more rigorous safety and environmental standards are set and followed. We know no single incident (such as a blowout) is produced by natural causes alone, but instead they all include the contribution of human mistakes, which should have been avoided or anticipated.

What we would like the readers of this book to understand is that we are citizens of the world before being energy industry leaders. We accept that our industry bears an extremely high responsibility and must continue to strive to improve every aspect of operations, as sustainability of our practices is a primary target for us even more than for others, including the industry's detractors. Our credibility rests on our being outspoken and honest in discussing these fundamental responsibilities. This book adopts that same philosophy, making it a potential bestseller and required reading for those who want to better understand the energy industry.

Jacques Bosio, *Paris, France*
Steve Holditch, *College Station, Texas*
Giovanni Paccaloni, *Milan, Italy*
Ed Runyan, *Midland, Texas*

June 2010

FOREWORD

The Energy Imperative is an important new publication that enters the market at a critical time in the aftermath of the *Deepwater Horizon* tragedy and environmental disaster. It is a time when the immediate trauma and spill response to the incident move us now to the public policy review and legal adjudication period where important decisions on the future of US energy solutions and environmental protections will set the energy course for the coming years. Lack of understanding surrounding our current energy system and the consequences of directionally pointing energy policy in new and different directions could have significant and long lasting deleterious impact on the availability, affordability and sustainability of our future energy system, depending on what is decided in the near future.

The Energy Imperative provides a frank, factual and understandable overview of the existing energy system in the country together with an assessment of the prospects for changes to the system that could provide greater diversity, some environmental improvements and potentially much higher costs to consumers. It is an affirmation of the efficacy of the past and current energy supply, demand, environmental and infrastructure build-out and operating system across the US. It also forewarns that changes to the system predicated on aspiration and/or idealism risk shrinking historic supplies and escalating costs while describing the impact future changes will realistically have, particularly on supply and the environment.

The text covers the science, technology, engineering and mathematics of the energy system with a sufficiently light touch so as not to turn away the anticipated non-technical readership while it nonetheless confirms the extraordinary investments in knowledge and application that are required to deliver energy to consumers. The audience for this book could range from the curious general public to knowledgeable professionals whose careers cover the waterfront of industries and services throughout the economy. Everyone who uses electrons and liquid fuels would benefit by reading this book. It should be mandatory reading for staff who work for or join energy companies because of the energy system depth, breadth, scope and scale it covers.

Critics of the book might suggest it has a hydrocarbon bias. In reality it simply states the obvious, pointing out that preference for other forms of energy does not eliminate or reduce the importance of powering our energy supply with affordable and available hydrocarbon energy. It also takes the time to describe cleaner ways to utilize such energy. The text does not demean alternative forms for energy; it puts them in perspective and demonstrates factually that supporters of anything but traditional energy sources simply cannot successfully power the economy without sustained investments in historic sources. This reality may turn off idealistic drivers of energy system change. But it is more important to supply energy than to support the egos of those who dream otherwise.

The Energy Imperative falls short on explaining what it would take to expand the role of nuclear energy and avoids the difficulties of resolving nuclear energy waste management. It discusses the environmental implications of hydrocarbon use but does not cover the risk management and safety systems that protect continued investment in future hydrocarbon development. It deals factually with the global warming issue and thus risks offending global warming enthusiasts who advocate tackling climate management as man's twenty-first century mandate. It soundly covers the reality of waste management, not its ideological cousin. The politics of energy and the public policies that are needed from multiple levels of government to address supply, demand, environment and infrastructure to assure our energy future are not within the scope of the book.

It's reassuring to know that we have plenty of energy and the ability to develop it, how we got started and what we have built for ourselves. It is also reassuring to know that once we get past our adolescent fears surrounding future use of hydrocarbons and their implications there are plenty of reasons for further developing existing domestic resources to secure our energy future. The two big bets the authors make on the future of natural gas and its potential liquefaction are reasonable and sound. *The Energy Imperative* makes useful reading and practical knowledge available for curious people from all walks of life. It helps calm the conversation and helps support continuity on the heels of the 2010 Gulf of Mexico disaster.

John Hofmeister
Founder and CEO, Citizens for Affordable Energy
Author: *Why We Hate the Oil Companies: Straight Talk from an Energy Insider* (Palgrave Macmillan 2010)
August 19, 2010

Mr. Hofmeister is a former president of Shell Oil Company.

INTRODUCTION

The Importance of Energy

Energy is the most important commodity. Without energy, there is no electricity, no heating, no manufacturing and no transportation. And without these, there is no modern life. Energy supply, energy consumption, and energy politics are driving the global political, economic, and social debate and will continue to do so for decades to come.

And yet few issues in modern history have generated more ideology-driven misinformation than energy. Very often the public debate about energy ignores facts and, especially, how difficult it is to change certain facts, both in terms of logistics and cost.

The problem is often the huge gap between the theoretical and the practical, the latter affected by logistical and economic considerations. What some people would find desirable to do, based on their view of how the world should be, is often confronted by the abysmally small odds of making it happen that way in the world that is. Thus, it's possible – but should not be – for public proclamations by governments and non-government groups to omit any disclosure of the required path and cost for a transition into their desired goal. Many imply that government and taxes should provide, but even then the magnitude is rarely revealed and taxes should provide the required funds, but even then, the magnitude is rarely revealed.

Using less energy is not an attractive option. Study after study shows the unambiguous relationship between energy consumption and the per capita adjusted gross domestic product (GDP), a good indicator of standard of living. Conservation is not what many would want to believe. Most people will conserve energy only when it does not materially change what they want to do. And much of what people do is either directly productive or uses products directly created by other people. Limiting energy use through forced conservation will reduce employment and lower the GDP. Furthermore, energy efficiency does not reduce energy consumption because people use energy-efficient appliances even more. For example, as air conditioners become more efficient, people build larger homes and offices.

Worldwide, over 85% of the energy consumption is from fossil fuels (coal, oil, and natural gas). Interestingly, thirty years ago, when world energy demand was only 60% of today's level, fossil fuels already accounted for almost 90% of the energy supply[1]. Forecasts for 2030 by the Energy Information Administration (EIA) of the US Department of Energy (DOE) suggest that this will not change very much for the next generation – with well over 80% of the total energy mix still coming from fossil fuels – in spite of all the current rhetoric about alternatives and in spite of predictions that the total energy demand will increase by 62%. The much-touted solar and wind energy, combined, are unlikely to contribute more than 1% of world energy demand by then[2]. Fossil fuels, discussed next, will continue to dominate the energy picture for the foreseeable future. The significance of the alternatives, to be discussed in later chapters, is greatly exaggerated today.

CHAPTER 1
What Are Fossil Fuels and Where Are They?

Chapter 1
What Are Fossil Fuels and Where Are They?

In the context of primary sources of energy, fossil fuels are naturally occurring hydrocarbons: natural gas, crude oil, and coal. For most of the twentieth century, these three sources have accounted for over 85% of the global energy supply. For the US energy mix, their contribution is similar, with oil providing about 40%. The predominance of hydrocarbons in the global energy portfolio is sure to remain for many decades to come. Even modest global economic growth will spur increases in the world energy demand, which will lead to substantial investment in expanded exploration and development of these resources.

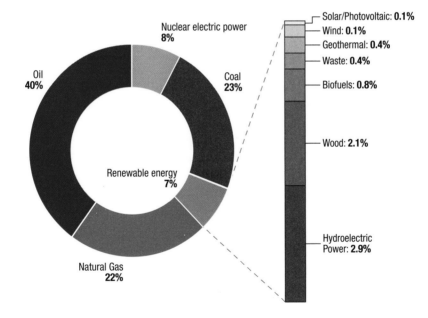

Figure 1.1. Contributions to Global Energy Supply (Source: EIA)

An obvious issue, and one that has shaped modern debate, is that naturally occurring hydrocarbons are not renewable and are subject to depletion. However, as we will discuss later in this book, this truism may take many decades if not centuries to emerge as a vital human issue but has already, on several occasions, caused premature alarm. These concerns have led to political pressure for changes, which in themselves are not backed with economic justification, nor are the proposed solutions economically viable. Alternative energy sources will eventually have to be developed and marshaled, but their evolution should not circumvent market forces, nor should they arise from ideological and/or political biases.

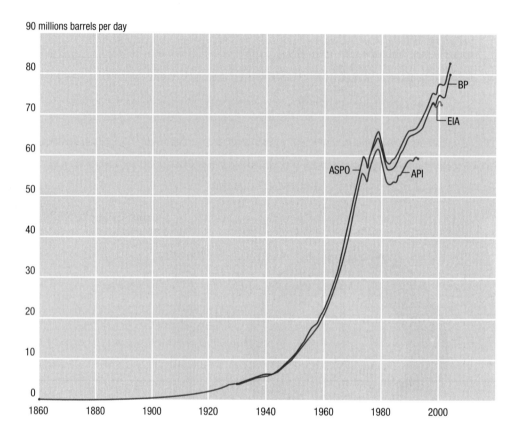

Figure 1.2. Global oil production since 1860
(Source: API, ASPO, BP, EIA)

The energy industry's evolution has been fast and imposing, considering where the industry started and how far it has come in the course of a little over a century. An example is the escalation of global oil production from very little prior to the twentieth century, to 10 million barrels per day as recently as 1950, to over 85 million barrels per day in 2010. As world energy demand has continued to increase, the industry has demonstrated remarkable creativity and the ability to always develop advanced technology to meet the challenge.

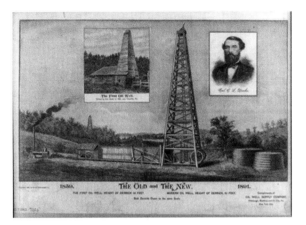

*Figure 1.3. Oil well drilled to a depth of 69 feet by
"Colonel" Edwin L. Drake, Titusville, Pennsylvania*

Especially noteworthy is the rate and magnitude of technology development and commercialization, from the reported first modern oil well drilled in 1848 near Baku, Azerbaijan, to the famous Drake discovery well of 1859 in Pennsylvania (Figure 1.3), to the first oil field (Figure 1.4), to wells drilled in the most remote locations on the globe: desert, jungle (Figure 1.5), permafrost (Figure 1.6), offshore, shelf to deepwater, drilling through sand and shale and salt and granite, to reservoirs with pressures of 20,000 psi and temperatures of 450 °F. From deep, sour, high-pressure gas and deep coal mines, to surface deposits of tar sands, to wells drilled more than 6 miles (nearly 10 kilometers) below the earth's surface, and wells drilled horizontally with entire horizontal sections extending through oil-bearing or gas-bearing rock sections 20 feet (6 meters) thick or less; even a well drilled in Qatar 40,320 feet long (12,293 meters), with a horizontal section of 35,770 feet (10,905 meters).[3]

Figure 1.4. First oilfield rapidly following the Drake discovery

Figure 1.5. Well drilled in the Taklamakan Desert, Xinjiang China

Figure 1.6. Drilling site in the arctic

Figure 1.7. Offshore platform, Norwegian sector of the North Sea

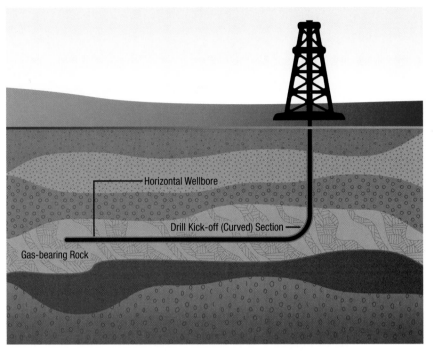

Figure 1.8. Depiction of horizontal well

And not forgetting multi-lateral wells and other technical accomplishments that early drillers could not even imagine.

The industry has performed countless tremendous feats of engineering and well architecture, the complex deepwater oil and gas platforms and production facilities with multiple wells through thousands of feet of water, drilled and completed to depths extending thousands of feet beneath the sea floor, often deviating from one another at high angles.

As late as 1950, there was virtually no offshore oil production. The first producing platform out of sight of land was installed off the coast of Louisiana by Kerr-McGee in 1947. By 1960, shallow (water depth of a few tens of feet) Gulf of Mexico production was established as a significant contributor to US oil production. Deepwater production, a definition that itself evolved from a few hundred feet to eventually 10,000 feet of water and more, did not begin for another 20 years. Deepwater natural gas production did not begin until the 1990s, but by 2010, is approaching 4 billion cubic feet of production per day (Figure 1.10).

Also, as late as 1995, deepwater oil and gas production as a percentage of total US production was small. But by 2010 it accounted for over 30% and is projected to exceed 35% (Figure 1.11).

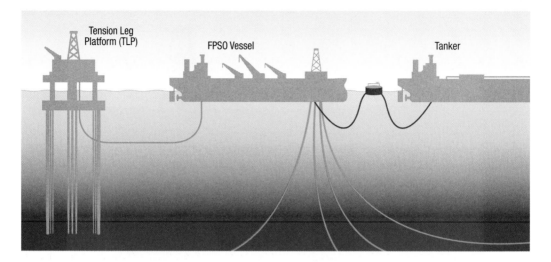

Figure 1.9. Rendition of Tension Leg Platform (TLP) with Floating Production, Storage and Offloading (FPSO) Vessel

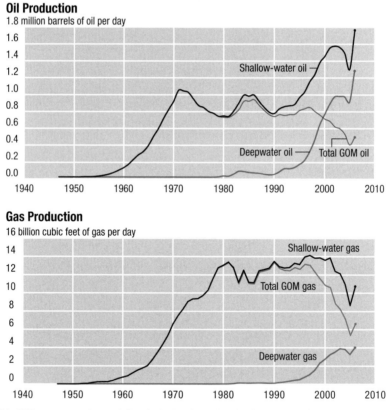

Note: 2006 average annual gas and oil production is estimated based on first 8 months of 2006.

Figure 1.10. Historic Gulf of Mexico (GOM) Shallow and Deepwater Oil and Gas Production (Source: Energy Tribune)

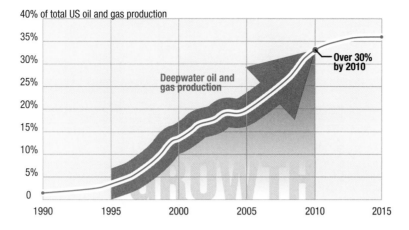

Figure 1.11. US deepwater oil and gas production – percent of total (Source: EIA)

There are low-cost wells that produce continuously for years (Figure 1.12), and there are extremely high-cost wells that never produce a drop of oil or a bubble of gas. There are also feats of chemical transformation (refining and natural gas processing), and of product distribution. From the first oil refineries (Figure 1.13) to the latest (Figure 1.14), these facilities can refine the heaviest of crude oils to gasoline, other fuels, finished products, and precursors to the seemingly endless number of petrochemical products that we completely rely on in all aspects of everyday life.

Figure 1.12. The remains of Well Harbell Farm No. 1, Four Mile Valley, California. Well produced continuously for 114 years. Shut in after 1 barrel of oil per week rate ceased.

Figure 1.13. Pioneer Oil Refinery of Star Oil Company. First commercial oil refinery in California (Source: California Office of Historic Preservation)

Figure 1.14. Section of a Modern Oil Refinery

Despite its amazing technical accomplishments and its crucial importance in supplying the world with fuel, power and raw materials, and in spite of an environmental record that is superior to that of many other industries (e.g., the paper or food processing and manufacturing), the energy industry has developed an unfavorable image in nearly all circles outside of its own, in both developed and undeveloped nations. Images such as oil-drenched sea birds, unclean working conditions and explosions are common when the media discusses the energy industry and tend to make a lasting impression in the minds of the general public. In April 2010, a deepwater blowout in BP's Macondo field in the Gulf of Mexico provided ample grist for the mill of public opinion. Explosions and fire killed 11 men and caused the sinking of Transocean's state-of-the-art drilling rig, Deepwater Horizon, which was contracted to BP. The loss of well control and failure of subsea wellhead equipment resulted in an initially estimated 5,000 barrels of oil per day (about 1,000 cubic meters per day) of crude oil leaking from the damaged well into the sea. These estimates were subsequently revised to flow rate levels five to ten times greater. The fact that this happened only 40 miles (64 kilometers) from the US Gulf Coast, in America's own backyard, caused an uproar. Many people questioned the need for offshore drilling and certainly the impression created by the media and politicians was of an industry with lax safety standards, prepared to cut corners in the interests of greater profits.

Figure 1.15. Bird drenched in oil from South Korea oil slick

Figure 1.16. Coal Miner in China

Figure 1.17. Iplom oil refinery of Busalla, near Genoa, northern Italy, Friday, Sept. 2, 2005

Because of such impressions, many people have come to believe that the industry's downsides (finite hydrocarbon reserves, pollution potential, gasoline prices, and the relatively recent climate change arguments) outweigh the considerable upsides (petroleum products; modern lifestyle and conveniences; health; the creation of individual, organizational, and national wealth; and on and on). Some of this perception stems from poor public relations efforts by the energy companies. It also indicates that most people have an imperfect understanding of how much their way of life depends on continued access to these vital resources for the foreseeable future. People simply are not aware of the dire consequences of not having adequate energy supply, and the enormity of the task to find alternatives. The problem is further compounded by irresponsible politicians and clouded by social engineers and ideologues.

Figures 1.18. and 1.19. The creation of wealth.
On the top, the magnificent Petronas Towers, Kuala Lumpur, Malaysia.
On the bottom, Dubai

There is also the perception, some of it the remnant of the colonial past, that major integrated oil companies (IOCs), "Big Oil," are reaping outrageous and unfair profits. However, it can be argued that in terms of net income as a percentage of sales, the oil and gas industry (shown as "oil and natural gas") is not exceptional among major industries: not at the bottom, but nowhere near the top (Figure 1.20).

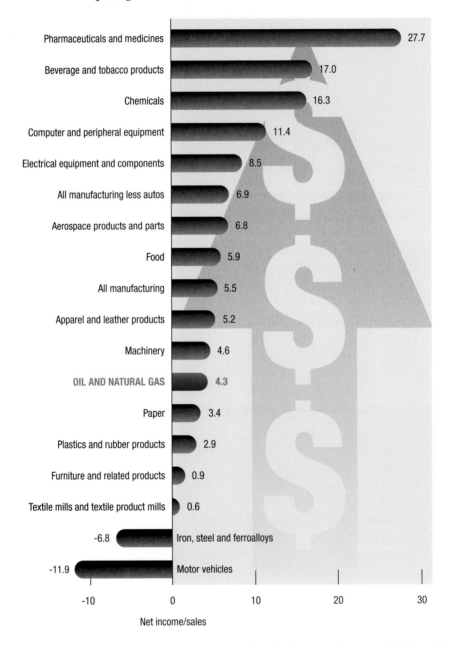

Figure 1.20. 2009 Comparative Earnings by Industry (Source: Oil Daily)

The entire global energy picture – from the beginning to the present day and on into the future – is connected by a single common thread: hydrocarbons. The ability to meet future energy demand is predicated on access to, and continuing utilization of, hydrocarbon resources. Hydrocarbons provide the fuel for the world, the fuel for the advancement of life, and the incentive for their replacement. This book endeavors to delve into the key aspects of this vital industry, now and in the future. Our aim is to provide the reader with a clear understanding of the energy business and an appreciation of the challenges we face collectively in the coming years.

But first – and before leaving Chapter 1 – what are hydrocarbons and where are they?

What are hydrocarbons?

Hydrocarbons are organic compounds constructed of hydrogen (H) and carbon (C) atoms, which can combine together along with smaller amounts of other atoms such as oxygen, nitrogen and sulfur to form an almost unlimited number of combinations. This characteristic of carbon in combination with hydrogen enables the multitude of chemical compounds that are listed to date – over 20 million! Plus, hydrocarbons can generally be mixed (that is, they are "miscible"), enabling uniquely complex mixtures such as crude oil.

Hydrocarbons are combustible fuels, and with oxygen they react according to the following:

$$\text{Hydrocarbon (Fuel)} + \text{Oxygen}$$
$$\rightarrow \text{Heat} + \text{Carbon Dioxide} + \text{Water (Vapor)}$$

The energy released in the form of heat – generated by the combustion reaction – is the key component. More recently, increasing environmental (and political) significance has been laid upon the associated by-product of the combustion reaction: carbon dioxide (CO_2). This is the gas blamed for so-called anthropogenic (man-made) global warming (AGW). Although many scientists and engineers disagree about the significance of CO_2 on earth's temperature and AGW has been dealt some serious blows, it is politically expedient to minimize the amount of carbon dioxide generated relative to the amount of hydrocarbon fuel (and oxygen) required to generate a particular amount of heat (energy).

Chemically, the simple combustion reaction for hydrocarbons is as follows:

$$C_x H_y + (x + 0.25y)O_2$$
$$\rightarrow \text{Heat} + xCO_2 + (0.5y)H_2O$$

Example hydrocarbon combustion reactions:

$$CH_4 (\text{methane in natural gas}) + 2O_2$$
$$\rightarrow \text{Heat} + CO_2 + 2H_2O$$

$$C_8H_{18} (\text{iso-octane in gasoline}) + 12.5O_2$$
$$\rightarrow \text{Heat} + 8CO_2 + 9H_2O$$

For each fuel, a heat of combustion (or heating value) can be measured. On a per weight basis, pure methane (CH_4), the simplest hydrocarbon, has a higher heating value than crude oil and gasoline (of which iso-octane is a component), which in turn have higher heating values than coal. Table 1.1 provides a comparison of example heating values for various fuels.

Table 1.1. Heats of Combustion (Heating Values) for Example Fuels

Fuel	Approximate Heating Value *(in megajoules per kilogram)*[1]
Hydrogen	142
Natural gas	55
Gasoline	47
Kerosene	46
Diesel	45
Ethanol	30
Methanol	23
Coal	15 – 35
Wood	12 – 15

Of associated importance – or at least current public focus – with respect to the primary hydrocarbon energy sources of natural gas, oil and coal, is the ratio of carbon (C) to hydrogen (H) weight. The highest "carbonization," in terms of the C:H ratio, is found in coal, a solid. In coals, C:H weight ratio exceeds 8, and they emit the greatest amount of CO_2 compared to other hydrocarbon fuels. Pure methane, the primary component of natural gas, has a C:H weight ratio of 3. Oil, with its wide variety of hydrocarbon components with larger molecular weights, falls between natural gas and coal in C:H weight ratio. Based on the combustion equation, methane will generate about 22% less CO_2 than iso-octane (and about 25% less than gasoline), given the same amount of oxygen. On the same basis, and very roughly speaking, natural gas emits about 30% less CO_2 than oil and approximately 50% less CO_2 than coal when creating energy.

While these values have relative meaning, they can be misleading, in that the capture, processing, storage and even re-use of CO_2 from the different sources will differ too (as will be discussed in later chapters). Also note that on an equivalent volume (rather than weight or mass) basis, hydrogen provides only 29% of the value of gasoline (more on this in Chapter 9).

Properties of Natural Gas

The principal constituent of natural gas is methane – the smallest and simplest of all hydrocarbons, made of one carbon atom and four hydrogen atoms (CH_4).

Natural gas is not purely methane but typically a mixture of methane plus other increasingly higher molecular weight hydrocarbon gases: ethane (C_2H_6), propane (C_3H_8) and butanes (C_4H_{10}); along with non-hydrocarbon impurities such as nitrogen (N_2), carbon dioxide (CO_2), and problematic hydrogen sulfide (H_2S) which must be removed before any natural gas is transported or sold. Natural gas may also contain rare gases, including helium (He). Natural gas is, in fact, the primary global source of helium. Low levels of mercury (Hg) may also appear in natural gas as an impurity. H_2S and Hg, if present, must be removed from natural gas before transporting.

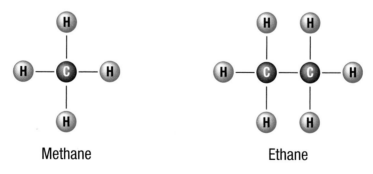

Methane   Ethane

Figure 1.21. Methane and Ethane Molecules

Natural gas is lighter than air and highly flammable. Because of its high H:C ratio, it is the cleanest burning of the hydrocarbon fuels, giving off less CO_2 and only minor amounts of pollutants such as sulfur dioxide (SO_2) and oxides of nitrogen. Natural gas can be classified as 'sweet' or 'sour', depending on H_2S content. Sweet gas has very low to no H_2S. Gas is considered sour if the H_2S content is above about 4 parts per million (ppm).

While compositions vary widely, Table 1.2, below, summarizes typical natural gas component concentration ranges.

Table 1.2. Typical Natural Gas Composition

Gas	Weight % range
Methane (CH_4)	70 – 90
Ethane (C_2H_6)	5 – 15
Propane (C_3H_8)	0 – 5
Butanes (C_4H_{10})	0 – 5
Carbon dioxide (CO_2)	0 – 8
Oxygen (O_2)	0 – 0.2
Nitrogen (N_2)	0 – 5
Hydrogen sulfide (H_2S)	0 – 5
Rare (Noble) Gases:	
Helium (He)	
Argon (A)	
Neon (Ne)	
Xenon (Xe)	

Properties of Oil

Oil (or petroleum, Greek for "rock oil"), is liquid. Before refining, produced "crude" oil is made up of larger (longer chain) and complex hydrocarbon structures, to varying degrees. Crude oil molecules contain five carbons (pentane, C_5H_{12}) or more. Generally, crude oil is comprised of hydrocarbons with 5 to 50-60 carbons. C_5 to C_{18} or C_{19} are generally liquids. The larger carbon-containing components are dissolved or suspended solids (depending on temperature). Hydrocarbons with less than five carbons are either natural gas components or natural gas liquids. The composite of the hydrocarbon components in oil

– their size, arrangement, and presence or absence of other non-hydrocarbon components such as sulfur or nitrogen incorporated into the liquid hydrocarbon structures – gives a particular crude oil its own unique chemical and physical (including flow) properties.

The physical classification of crude oil is generally based on its API (American Petroleum Institute) gravity, which is a measurement of oil density relative to water. Oil ranges from light to intermediate to heavy to nearly solid 'extra heavy' crudes. The corresponding API gravity of such oils ranges from above 30 (light) through the 20s (medium) to the teens (heavy) and into the single digits (extra heavy oil). Crude oils with higher API gravities (lighter oils) are typically more valuable. Appendix B includes more details about crude oil classifications.

Oil is also categorized by its viscosity, which generally correlates to gravity: the lower the gravity (the heavier the oil), the higher the viscosity. There are a great variety of crudes and grades, as properties are dependent on the crude oil's place of origin – the ground and conditions under which it was formed. And as mentioned above, crude oils are categorized as 'sweet' or 'sour', depending on their sulfur content. All contribute to the greater energy picture, with about 84% of crude oil, by volume, used for fuels.

Heavy oil, simply put, is oil that does not readily flow from the well to the surface. Heavy oil, and its recovery, is becoming increasingly important strategically and crucial to meeting future energy demand growth. Probably 70% of all presently recoverable heavy oil reserves are in the Western Hemisphere. Most is in South America (primarily Venezuela), followed by North America (including the tar sands of Canada). Total heavy oil reserves exceed those of medium and light oil combined (see Chapter 2). However in 2010, heavy oil is still a minor contributor to world oil production. Heavy oil mostly requires mining (tar sands) or some form of thermal stimulation (injection of steam and/or superheated water to lower the viscosity of oil and induce flow). Most operators have not been able to produce heavy oil economically at prevailing oil prices.

With the inevitable advancement of heavy oil recovery technology and methodologies, however, it is only a matter of time before exploitation of heavy oil becomes much more common.

The chemical composition of crude oil, in general, varies according to where it was obtained. As mentioned previously, this largely has to do with the type of ground in which the oil was formed, and what contaminants were present and in what relative concentrations. It is then through the refining processes that the hydrocarbons native to each crude oil are separated, processed and "reformed" to different synthetic hydrocarbon compounds, leading to the broad range of petroleum products – including commercial fuels. Appendix C contains the classification and characteristics of the hydrocarbons in crude oils.

Properties of Coal

Coal is a solid hydrocarbon fuel, used to produce electricity and heat through combustion. Coal contains an even greater number of carbon atoms than oil, and a much lower hydrogen-to-carbon ratio. Coal is a hard, black solid, containing carbon, hydrogen, oxygen, nitrogen, and varying amounts of sulfur. There are three basic types of coal: anthracite, bituminous and lignite. Anthracite coal is the hardest variety, having most carbon. Because of the high carbon content, anthracite also provides greater energy content than the other coal types. Lignite is the softest coal type, as it is lowest in carbon and relatively high in hydrogen and oxygen. Bituminous coal is in between the two in its physical properties.

Figure 1.22. Coal

Of all forms of coal, anthracite has the highest carbon content (about 86 to 98%) and highest heat value (about 30 to 35 MJ/kg). Carbon content in bituminous coals ranges from about 46 to 86%, and in lignite, 46 to 60%. As carbon content decreases, so does the heat value. Impurities such as sulfur and nitrogen are also significant in that they also reduce heat value and have unfavorable environmental implications.

These combustible hydrocarbons – oil, natural gas and coal – are also known as fossil fuels. *Fossil* is an association to their period of origin. While there is some mystery to the exact origins of the fossil fuels, the prevailing theory is the Biogenic

Theory. This theory holds that the origins of oil and natural gas spanned a period of perhaps 400 million years, from the Cambrian up to the Jurassic geologic periods. The primary period of origin is postulated to have been during the Carboniferous Period, over 300 million years ago, and spanning about 75 million years. During that time, the climate was such that plant growth proliferated far more than today. Small to large plant life, as well as very tiny marine plant and animal life , called phytoplankton and zooplankton, respectively, predominated. As these lifeforms died, they dropped to the bottom of the swamps and oceans in which they thrived. Over time, they were buried deeper and deeper in layers of sediment – and as their burial deepened, the pressure and temperature increased. Gradually, over millions of years, the organic matter from these long-dead organisms was "cooked" and transformed into the various fossil fuels that are partially accessible today. Natural gas is likely to be the result of longer exposure to higher pressure and temperature, light crude oil is next, heavy oil follows and finally coal. This also means that deep reservoirs are likely to contain mostly gas, whereas heavy oil is found in relatively shallow geological structures. Coal is often located near the surface and is accessed through mining.

Oil and natural gas are accessible through wells drilled into the underground rock structures, or "reservoirs", in which they are contained. One common misconception is that oil and gas are present in underground "lakes", and the wells drilled are the "straws" that enter into and suck out the hydrocarbons. This is not at all the case. Oil and gas are "trapped" in the tiny pore spaces and sometimes natural fractures present in sandstone, carbonate, and shale rock formations – not unlike how water is held by a sponge. The weight of all the rock layers above squeezes the reservoir rock and pressurizes the fluids contained in the pore spaces. Oil and gas are produced by relieving or "drawing down" this natural pressure via the well. However, relying on this natural pressure alone would result in much of the oil and gas being permanently left behind. So, various techniques and new technologies are used to improve the proportion of hydrocarbon that can be extracted from the pore spaces. The portion that can be produced is referred to as "recoverable" and it can range from less than 10% to over 50% of the hydrocarbons in place in the reservoir.

The presence of certain hydrocarbon chemical structures in oils and coals, as well as impurities (e.g., sulfur, nitrogen), are indicators of their biological origin.

An alternative to the Biogenic Theory, at least for oil and gas, is the Abiogenic Theory. This theory was originally championed in the western world by astronomer Thomas Gold, based primarily on the studies of Russian Nikolai Kudryavtsev, 250 years ago. The theory posits that oil forms naturally, deep below the surface of the Earth, and seeps upward through fractures. Further suggestion has it that the fracture channels were formed by collisions of asteroids with the Earth.

The Abiogenic Theory is not taken seriously in the petroleum industry and it is not a basis for exploration. The presence of abiogenic oil is not necessarily refuted, but the quantity and its contribution to worldwide hydrocarbon reserves is generally dismissed as insignificant.

For all practical purposes, hydrocarbon fuel resources are considered to be biogenic and thus finite. Petroleum geology explains where they are located beneath the earth's land surfaces and subsea. It can explain why one area may contain vast resources of oil or gas or coal – while a nearby area or country may contain none.

Where are the Hydrocarbons?

Natural Gas

Sometime between 6000 and 2000 years BC, the first discoveries of natural gas seeps were made in ancient Persia, today's Iran, and likely contributed to the worship of fire practiced by the Persians. Now, recoverable natural gas is found in oil fields ("associated gas") or in isolated gas fields, and from increasingly challenging resources as discoveries and production technologies advance – gas-bearing shales, low-permeability rock formations ("tight gas"), and coal beds ("coalbed methane"). The world's largest proven gas reserves are in Russia. Following Russia in greatest proven reserves are Iran, Qatar, Saudi Arabia, and the United Arab Emirates. The world's largest gas field is Qatar's offshore North Field, estimated to contain 900 trillion cubic feet (25 trillion cubic meters) of gas in place. That alone would be sufficient to last more than 200 years at properly managed production levels.

Oil

Oil has been used for more than 5,000 years. In his writings, Herodotus observed the use of asphalt ("pitch") in construction in Babylon. The material was collected from seeps at Tuttul (modern-day Hit) on the Euphrates River, and by the ancient Sumerians and Assyrians. The first oil wells are believed to have been drilled in China in the 4th century or before, using crude bits attached to bamboo poles.

Needless to say, the history of oil discovery and development is a long, fascinating, and far-reaching story.

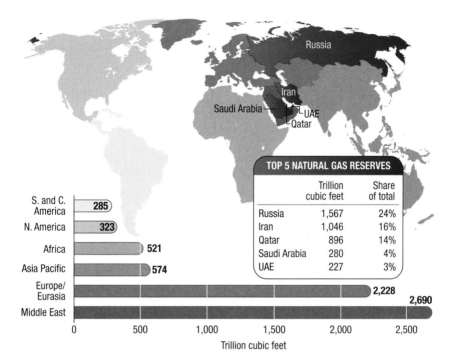

Figure 1.23. Proved reserves in trillion cubic feet – end of 2008
(Source: BP Statistical Review 2009)

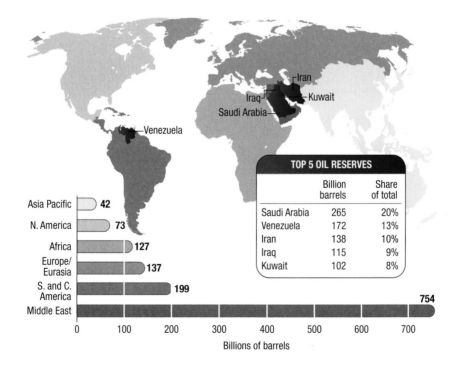

Figure 1.24. World Recoverable Oil Reserves (Source: BP Statistical Review 2009)

Figure 1.24 provides an "oil map" – indicating by relative scale the greater contributing regions to global (recoverable) oil reserves today.

Coal

The earliest known use of coal as a fuel is believed to have been in China, as early as 3,000 years ago. It is believed to have been used to smelt copper – with the "stone that could burn." World proven and producible coal reserves alone are estimated to be more than 900 billion tons.[4] At the current coal production rate, this would last over 160 years, without adding any new reserves. However, new drilling and mining programs continually add reserves.

Of the three fossil fuels, coal has the most widely distributed reserves. It is mined in over 100 countries and on every continent except Antarctica. The largest reserves are in the United States, Russia, Australia, China, India and South Africa.Table 1.3 provides estimated coal reserves by those countries collectively holding 90% of world coal reserves.

Figure 1.25 summarizes an estimate of the amounts of oil equivalent contributed by natural gas, oil, and coal by global region.

Table 1.3. Proved Recoverable Coal Reserves In Million Metric Tons: 90% of World Coal Reserves in 8 Countries (Source: BP Statistical Review 2010)

Million Metric Tons	Anthracite and Bituminus	Sub-bituminous and Lignite	Total	Share of Total
US	108950	129358	238308	28.9%
Russian Federation	49088	107922	157010	19.0%
China	62200	52300	114500	13.9%
Australia	36800	39400	76200	9.2%
India	54,000	4,600	58,600	7.1%
Ukraine	15,351	18,522	33,873	4.1%
Kazakhstan	28,170	3,130	31,300	3.8%
South Africa	30,408	-	30,408	3.7%

It is true that the naturally occurring underground hydrocarbon resources – natural gas, oil, and coal – have a finite life. How long? That is the subject of continuing controversy. However, make no mistake, there are enormous reserves of each, especially when considering not only those that are recoverable now, but those that will also become recoverable as new technology is developed in the future. Thus, although our society is, today, fueled almost exclusively by hydrocarbons, we will progressively transition to a society that grows fuel crops and harvests energy from other sources. However, this change will happen only very gradually, with slowly increasing contributions from alternative, renewable fuels and renewable sources.

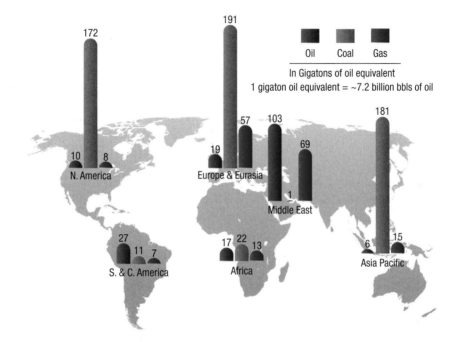

Figure 1.25. Fossil fuel reserves in gigatons of oil equivalent
(Source: BP Statistical Review 2009)

Given that hydrocarbons remain so important, the next questions are who controls the current oil and natural gas reserves and production of these hydrocarbons? Who wants them? Who uses them? These are addressed next.

CHAPTER 2
Who Controls the Oil and Gas?

Chapter 2
Who Controls the Oil and Gas?

Ever since the formation of the Standard Oil Trust in 1863 by John D. Rockefeller, and its monopolistic control of the great majority of production, transport, refining, and marketing of oil and petroleum products, "who controls the oil" has always been viewed with fascination and disdain. There is no more identifiable energy industry sector than that of the major investor-owned international oil companies (IOC) or as they are more often called now, the "Super Majors." The emergence of the IOCs began with the government-imposed break-up of the Standard Oil Trust in 1911, which spawned the many different Standard Oil companies, and the growth of each to much greater value than the original company. Additionally, companies such as British Petroleum and the Royal Dutch/Shell Group emerged. BP started as the Anglo-Persian Oil Company in 1909. The Royal Dutch/Shell Group was formed in 1907 through the merger of the Royal Dutch Petroleum Company and the Shell Transport and Trading Company of the United Kingdom – both of which were founded in the 1890s. The merger was struck to compete with the dominant American oil industry that was rapidly expanding globally.

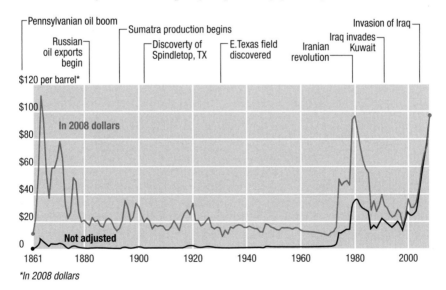

Figure 2.1. Oil Price History – 1861-2008
(Source: BP Statistical Review 2009)

The twentieth and twenty-first centuries have seen a multitude of friendly and unfriendly oil company "takeovers," seemingly endless mergers and acquisitions by (and of) small and large companies, spinoffs, and re-mergers. The investment opportunities and implications have been spectacular and thus have garnered considerable public attention. The oil and gasoline price hikes and spikes between 2000 and 2008 caused anguish and, even if geopolitically caused or motivated, invariably generated bad public outcry against "villainous" Big Oil. There will be more of that misplaced blame to come too, as oil prices will rise substantially again with the unrelenting demand for oil, led especially by China and India.

The Seven Sisters

The original Seven Sisters were the major oil companies that dominated world markets after World War II. The name was coined by Enrico Mattei, perhaps the most acclaimed entrepreneur in post-World War II Italy.

The sisters were the three companies formed by the breakup of Rockefeller's Standard Oil plus four other major oil companies:

1. Standard Oil of New Jersey *(later Exxon)*
2. Standard Oil of New York *(later Mobil and eventually merged with Exxon)*
3. Standard Oil of California *(later Chevron)*
4. Anglo-Persian Oil Company *(later the Anglo-Iranian Oil Company, then British Petroleum in 1954, then BP-Amoco after merging in 1998, and renamed BP in 2000)*
5. Royal Dutch Shell *(formed in the 1907 merger of Royal Dutch Petroleum Company and the Shell Transport and Trading Company, Ltd., UK)*
6. Gulf Oil *(acquired primarily by and absorbed into Chevron in 1985)*
7. Texaco *(merged into Chevron in 2001)*

These Seven Sisters, at the time, organized themselves as a cartel in order to effectively negotiate internationally, especially with Third World operators in the developing world. However, in time, the control that the Seven Sisters imposed began to decline, starting with the actions of Enrico Mattei. Mattei was an anti-fascist partisan and leader during World War II. The new Christian Democrat government that took power in Italy after the war appointed him as special commissary for AGIP, the Italian Petroleum Agency created by the Fascist government. His role was to liquidate assets. However, he did the opposite, instead re-establishing and growing the organization and eventually, by 1953, creating the National Fuel Trust, Ente Nazionale Idrocarburi (ENI). He established oil concessions in the Middle East and a trade agreement with the Soviet Union, both of which began the process of chipping away at the power of the Seven Sisters. Mattei died in a plane crash in 1962. Today, ENI is the sixth largest oil company in the world.

Figure 2.2. Enrico Mattei (1906 – 1962)

The collective power and influence of the Seven Sisters began to decline more significantly once OPEC (Organization of the Petroleum Exporting Countries) was formed in 1960, and especially into the 1970s, as the Arab states were able to exercise their control over oil prices and production. Remaining today of the original Seven Sisters are:

1. ExxonMobil *(merger of Exxon and Mobil in 1999)*
2. Chevron
3. BP
4. Royal Dutch Shell

These, plus Total S.A. of France (formed with the mergers of Total and Belgian Petrofina in 1999 and then with Elf Aquitaine in 2000), and ConocoPhillips (formed by the 2002 merger of Conoco, formerly Continental Oil Company, and Phillips Petroleum), comprise the current "Super Majors" – or so-called "Big Oil" – the largest of the non-state-owned oil concerns.

OPEC

The Organization of the Petroleum Exporting Countries (OPEC) is a cartel that was created at the Baghdad Conference on September 10–14, 1960, by Iran, Iraq, Kuwait, Saudi Arabia and Venezuela. The five founding members have been joined over the years by nine other members, two of which have suspended membership (Indonesia in 2009 and Gabon in 1994). Indonesia became a net importer in 2008 and thus withdrew, but may return if it becomes a net exporter again. Today, OPEC is a 12-member cartel, comprised of the following members:

Saudi Arabia *(1960)*	Libya *(1962)*
Iran *(1960)*	United Arab Emirates *(1967)*
Iraq *(1960)*	Algeria *(1969)*
Kuwait *(1960)*	Nigeria *(1971)*
Venezuela *(1960)*	Ecuador *(1973-1992; re-joined 2007)*
Qatar *(1961)*	Angola *(2007)*

One of the primary objectives of the OPEC cartel is to protect the individual and collective interests of the member countries – via oil price stabilization in the international markets. In recent years, OPEC control of, or influence on, the price of oil has not been as strong as in the past due to oil discoveries in the North Sea, the Gulf of Mexico and Alaska, and the opening of Russia and former Soviet Union nations and their large reserve base and daily production.

Figure 2.3 shows the fluctuation of OPEC to non-OPEC oil and petroleum product importation to the US, and the indication of generally reduced reliance.

Also, with respect to the US, Canada, not Saudi Arabia, is now its largest petroleum supplier (see Figure 2.4).

Still, OPEC nations account for about 30% of US oil supply (not including petroleum products). This amount of OPEC supplied crude oil is nearly the same as that produced domestically in the US. OPEC also holds two-thirds of global oil reserves and accounts for about one-third of global oil production. Therefore, despite diminishing reputation with respect to its power and control, OPEC's influence as a cartel in global oil markets should not be underestimated.

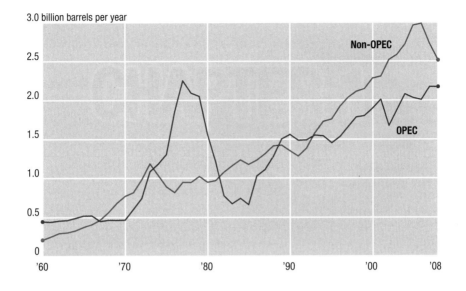

Figure 2.3. OPEC vs. non-OPEC oil importation to the US (Source: EIA)

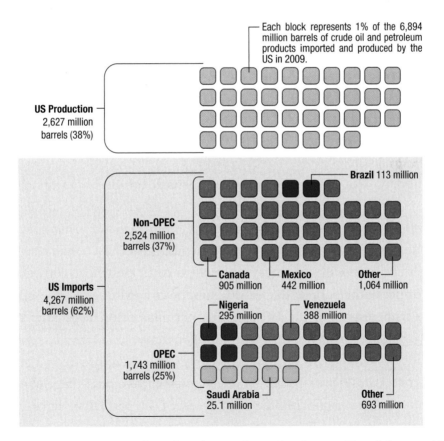

Figure 2.4. 2009 US Crude Oil and Petroleum Products – Breakdown of US Production vs. OPEC vs. Non-OPEC. (Source: EIA)

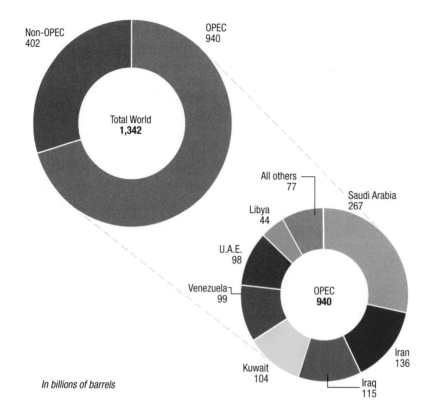

In billions of barrels

Figure 2.5. OPEC and non-OPEC Proven Oil Reserves (Source: Fuelwise.ie)

Natural Gas

Public attention has always been mostly focused on oil. What about natural gas? The US is substantially more self-sufficient when it comes to natural gas, importing only 16% of its needs, primarily from Canada. This compares to about 60% importation of oil from foreign countries. The US has also recently supplanted Russia as the number one natural gas producer. However, total production from the rest of the world, collectively (Middle East, Asia-Pacific, Africa), is rising more sharply, and the relative contribution to global production from the US (and North America) is trending lower.

When it comes to control of natural gas reserves, there has been a significant shift in recent years. Proved North American reserves, as a percentage of the global total, have declined while Middle East reserves, in particular, have increased dramatically (Figure 2.7). European and Eurasian natural gas reserves have also declined somewhat, but the collective amount controlled is still substantial – in large part due to the enormous gas reserves of Russia.

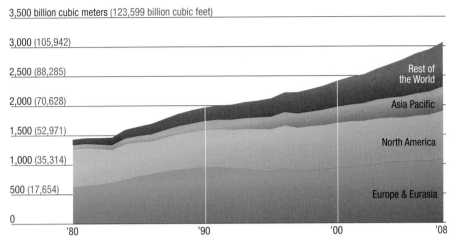

3,500 billion cubic meters (123,599 billion cubic feet)

3,000 (105,942)

2,500 (88,285)

Rest of the World

2,000 (70,628)

Asia Pacific

1,500 (52,971)

North America

1,000 (35,314)

500 (17,654)

Europe & Eurasia

0

'80   '90   '00   '08

Figure 2.6. World natural gas production by area – 1983 – 2008
(Source: BP Statistical Review 2009)

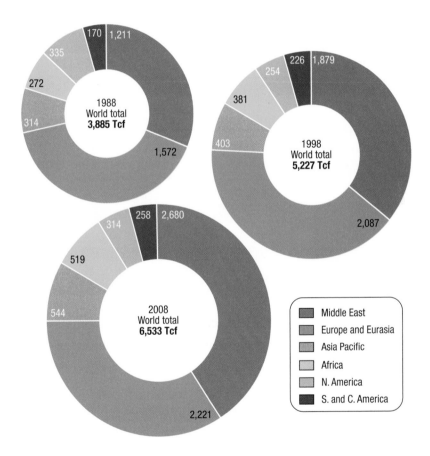

Figure 2.7. Distribution of Proved Natural Gas Reserves By Area – 1988, 1998,
2008 (Source: BP Statistical Review 2009)

Table 2.1. World Natural Gas Reserves by Country (Source: Oil & Gas Journal)

Country	Proved Reserves (Trillion Cubic Feet)	Percent of World Total
World	6,254	100.0
Top 20 Countries	5,674	90.7
Russia	1,680	26.9
Iran	992	15.9
Qatar	892	14.3
Saudi Arabia	258	4.1
United States	238	3.8
United Arab Emirates	214	3.4
Nigeria	184	2.9
Venezuela	171	2.7
Algeria	159	2.5
Iraq	112	1.8
Indonesia	106	1.7
Turkmenistan	94	1.5
Kazakhstan	85	1.4
Malaysia	83	1.3
Norway	82	1.3
China	80	1.3
Uzbekistan	65	1.0
Kuwait	63	1.0
Egypt	59	0.9
Canada	58	0.9
Rest of the World	581	9.3

Also quite striking are the differences in natural gas reserves-to-production (R/P) ratios for different global regions, as shown in Figure 2.8. In the US, for example, the natural gas reserve base has grown but the exploitation of those reserves has also grown. Therefore, the R/P ratio for the US is relatively small compared to that of other regions. In the Middle East, where significant production of the region's enormous reserves has only recently begun, in earnest, the R/P ratio is extremely high. Figure 2.8 presents R/P ratios based on reported proved natural-gas reserves.

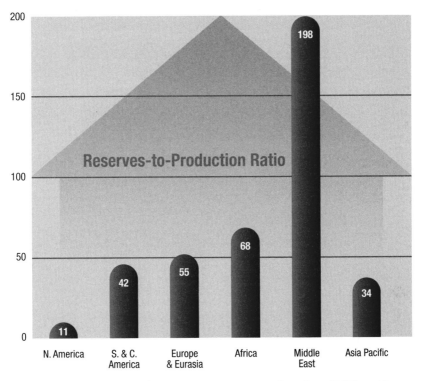

Figure 2.8. Natural gas reserves-to-production (R/P) ratios
(Source: BP Statistical Review 2009)

The New Seven Sisters

In March 2007, The Financial Times identified the "New Seven Sisters," the largest of the state-owned national oil companies (NOC) – those having the greatest control of oil and gas reserves, collectively, as well as production. They are:

1. Saudi Aramco *(Saudi Arabia)*
2. JSC Gazprom *(Russia)*
3. CNPC *(China)*
4. NIOC *(Iran)*
5. PDVSA *(Venezuela)*
6. Petrobras *(Brazil)*
7. Petronas *(Malaysia)*

Figures 2.9 – 2.11 indicate the collective dominance of the NOCs and their respective countries of origin, in terms of global hydrocarbon production, exportation, and reserves.

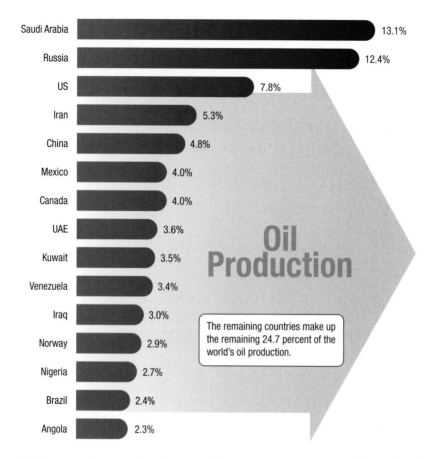

Figure 2.9. Largest oil producing countries as percent of world production (Source: BP Statistical Review 2009)

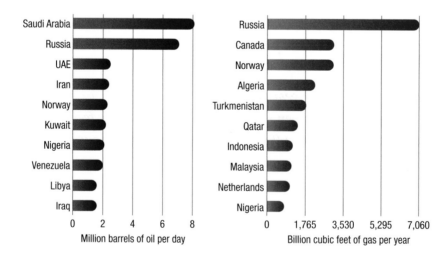

Figure 2.10. Largest oil (includes NGLs and condensate) and gas exporters – 2007 (Source: KBC Market Services)

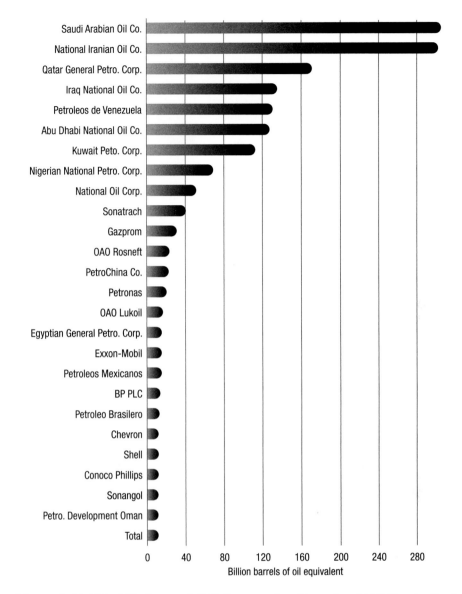

Figure 2.11. World's Largest Oil Companies: Barrels of Oil Equivalent (Oil and Natural Gas) Proven Reserves (every ~6,000 cubic feet of gas is one barrel of oil equivalent) (Source: (Source: Oil & Gas Journal, EIA)

The Myth of "Big Oil"

The energy world has changed dramatically since the reign of the original Seven Sisters. However, the general public still assumes that it is the large IOCs – the Super Majors or Big Oil giants ExxonMobil, Chevron, Shell, BP, etc. – that control oil prices as well as the majority of oil and natural gas reserves. Figures 2.12 and 2.13 show this is not at all the case.

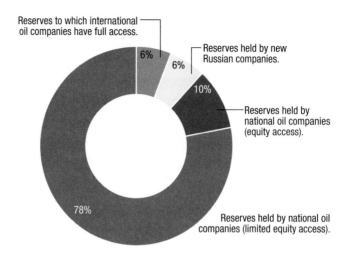
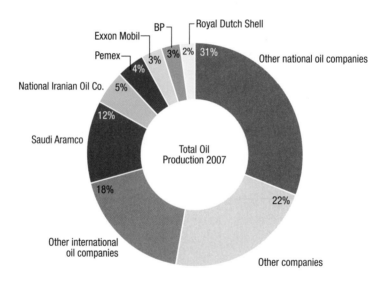

Figure 2.12. IOC access to oil reserves – relative to NOCs
(Source: PFC Energy, 2008)

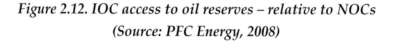

Figure 2.13. Total oil production, belying the myth of
"Big Oil" control (Source: Petroleum Intellegence Weekly)

Besides the fact that nearly 80% of world oil reserves are owned exclusively by NOCs, the contribution to world oil production by "Big Oil" pales in comparison to that of the NOCs.

In an energy world increasingly dominated by the NOCs, the Super Majors have focused their attention on growth through exploration in extreme locations such as those beneath deep and ultra-deep waters in the Gulf of Mexico. They are otherwise thwarted through legislation in the US, for example, to explore and develop remaining onshore and offshore "frontiers." They also increasingly focus on growth and expansion through acquisition of existing assets, especially those of the growing independent oil and gas producers. Through such "mergers," the Super Majors intend to operate with greater capital and efficiencies, and to increase their project and technology portfolios to what are currently referred to as "unconventional" oil and natural gas assets. One category of unconventional oil and natural gas assets is based on the permeability of the reservoir containing the oil or gas. Permeability is a measure of a rock's ability to contain and transmit fluid (oil, gas, or water). When a well is drilled into a rock layer containing fluid, and in which the pore spaces or natural fractures are well connected (i.e. high permeability), then fluid will flow naturally or more readily to the wellbore. If the permeability of the rock layer is very low, natural fluid flow will be very limited or non-existent.

Unconventional resources include those that can be produced from very low permeability rock formations – a challenge that requires specialized technology (see Chapter 4). Conventional reservoirs have higher permeabilities that enable oil or gas to flow more readily. Examples of "unconventional" resources are "tight gas" (gas held in ultra-low permeability rock formations, typically sandstones), shale gas, shale oil, and coalbed methane (CBM).

Heavy and extra heavy oil are also referred to as "unconventional" resources, not based on permeability of the rock holding the oil in place, but on the physical properties (high density and hydrocarbon chemical structure) of the heavy and extra heavy oils themselves, relative to more common (light and medium) crude oils (see Appendix B).

Shale Gas

The North American natural gas supply makeup has been shifting dramatically from conventional to unconventional gas; in particular, shale gas is of greatest future significance in North America. It is projected that unconventional natural gas will contribute over 50 percent of total natural gas production in North America by 2020.[5] Shale gas plays will be the major contributors, with today's total level of production approximately tripling in that time.

Of the surprisingly many gas and oil shales in the US, six are expected to contribute the lion's share of natural gas production – the Barnett (currently the

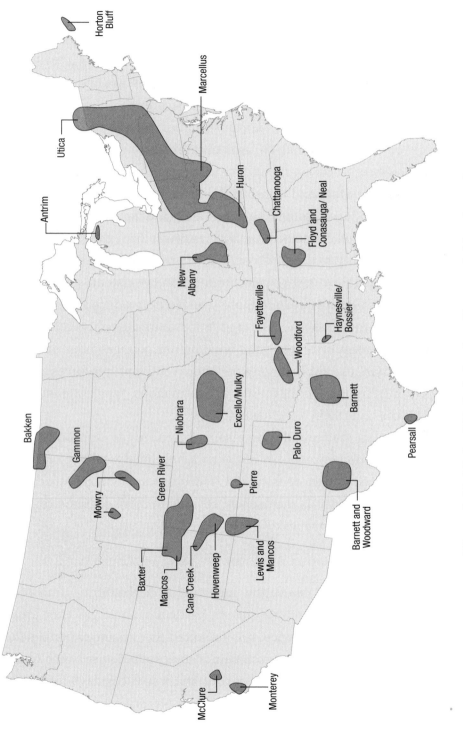

Figure 2.14. Major US shale basins (Source: Schlumberger)

largest), Antrim, Haynesville (potentially the greatest), Fayetteville, Woodford, and Marcellus (significantly growing activity). The Eagle Ford shale of South Texas (not shown above) is a more recent and promising entry, and could step up to round out a magnificent seven.

In Canada, exploitation of the Horn River shale is expected to ramp up. With this, the US (and North America) reserve base will continue to grow, along with total production. This is without mention of the Bakken shale (North Dakota, Montana, and Canada), a rare tight (low permeability) oil play which is poised for a substantial increase in drilling and completion investment and activity in the next several years.

These gas (and oil) shale developments have been achieved not by the Super Majors, but by the smaller, less-integrated independents and by ever-improving horizontal well drilling and hydraulic fracture stimulation technologies and methods (see Chapter 4). The top US producers of shale gas (and natural gas, in general) as of 2009 include XTO, Devon Energy, Encana, EOG Resources, Chesapeake, Southwestern Energy, Newfield, and a number of others. The rise of other shale gas plays and excitement in the associated communities will also no doubt emerge. Additionally, there will be new contributions to North American unconventional natural gas reserves and production from CBM, increasingly from, low-permeability "tight sands" (TS) (which provide the greatest share of current gas production), ultra-deepwater gas and shallow-water, ultra-deep (SWUD) developments.

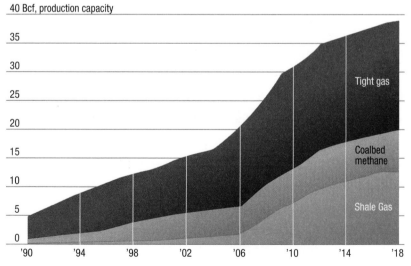

Figure 2.15. Projected US natural gas production from unconventional resources (Source: All Consulting)

Figure 2.16 shows the upward trend in new well contribution to total US gas production (lower 48 states) – thanks to shale gas, tight gas, and offshore developments.

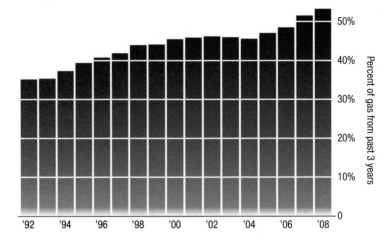

Figure 2.16. Recently drilled well contribution to US gas: 1992 – 2008
(Source: Bernstein Research)

The future of unconventional natural gas, especially shale gas, is certainly not confined to the United States and Canada. There may be nearly 700 shales in over 140 petroleum basins spread across the globe. Significant and more immediately potential gas shale plays exist in northern, central, and eastern Europe, China, Australia, and India, as examples. The Super Majors, as well as large independents, are investing in such shale plays overseas. Companies such as Total of France and StatoilHydro of Norway are also investing in US shale plays (both in partnership with Chesapeake Energy) in the Barnett shale of east Texas and the Marcellus shale in the northeast, respectively.

In November 2009, US President Barrack Obama indicated a desire to share US gas shale technology with China in the interest of US investment in the development of Chinese gas shales. Gazprom of Russia has made no secret of its interest in buying a US shale gas producer to learn what is necessary to exploit Russian shale gas opportunities. Potential shale gas opportunities are also present in northeastern France, northern Europe (Alum Shale), Germany and the Netherlands. In fact, ExxonMobil has invested in substantial acreage in the Lower Saxony Basin of Germany, as well as in the shale gas-containing Makó Trough in Hungary. ConocoPhillips is placing attention and investment in Poland. The following release from the Dow Jones Newswire on December 14, 2009 speaks for itself:

*Exxon Mobil Corp.'s (XOM) $31 billion bid for XTO Energy Inc. (XTO) marks the biggest endorsement yet for the exploitation of unconventional natural gas resources--and signals a growing will to carry the expertise that has revolutionized gas production beyond the North American gas market. In a statement released Monday, Exxon Chief Executive Rex Tillerson said that the deal would benefit consumers **"here and around the world,"** and help the company develop natural gas and oil resources globally. Exxon, the world's largest publicly traded oil company, has recently scooped up unconventional gas assets in Poland, Germany, Hungary and Argentina. If brought online, these resources could help feed major markets that currently face unsteady supplies.*

Heavy Oil

Heavy and extra heavy oil is another unconventional hydrocarbon source that is present in tremendous, largely untapped, abundance, and one that may shift the oil supply picture in coming years. Russia, the US, Canada, and some countries in the Middle East hold heavy oil deposits (i.e., oil sands) – in addition to Venezuela. Canada and Venezuela are the two largest holders and producers of heavy oil. Reserves of heavy oil in Canada and Venezuela alone account for nearly the same amount as that held worldwide in conventional oil. When accounting for heavy and extra heavy oil reserves, including Canadian tar sands, the total reserves of such unconventional oil is roughly twice that of present conventional oil reserves worldwide.

Finally, the Super Majors (and other major oil companies) are attentive to opportunities to invest jointly with the NOCs, as indicated in Figure 2.12 (10% of reserves held by NOCs). This will fuel the further creation and growth of the "International NOCs" but will also expand the international footprint of both the Super Majors and the large independents that can do business with NOCs. The NOCs do, after all, require access to technologies to aid in recovering their substantial reserves, and one way to accelerate that access is through collaboration with the Super Majors. However, the need for cash investment and technical support to the NOCs is not as great as it once was. When oil and gas prices are high, the NOCs are not as inclined to seek outside collaboration. And unlike the glory days of the original Seven Sisters, pricing – and thus future growth – can no longer be controlled by the Super Majors.

CHAPTER 3

Energy Use – Reality

Chapter 3
Energy Use – Reality

The connection between wealth and energy consumption is not really something to be debated or doubted. Many before us have shown the unambiguous correlation between the two. The first two figures in this chapter show how the 15 largest world economies fare in the correlation between energy consumption and wealth and oil consumption and wealth.

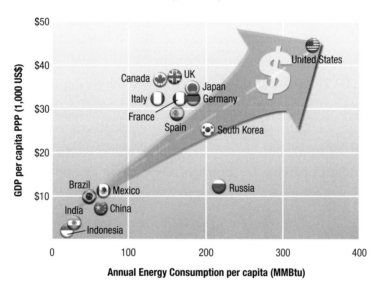

World Energy Consumption and Wealth

Figure 3.1. Gross domestic product per capita vs. energy consumption per capita (Source: EIA, CIA Fact Book, Time Almanac)

As can be seen in Figure 3.1, the United States is the richest nation in the world in purchasing power parity (PPP) gross domestic product (GDP) per capita; the country is also the biggest energy consumer. It should not be surprising: a rich country will be a bigger user of energy. Countries significantly above the correlation in Figure 3.1 are considered by some as more efficient. Those below are less efficient. Of course other factors influence these relationships, such as the size of the country, the overall development of the country and how its development and wealth are distributed. Russia, with abundant energy sources but with a resistance to changing its social and economic system, is significantly below the correlation. The country uses at least 25 percent more natural gas per capita than the United States. The gas is highly subsidized. China, undergoing world-shattering changes, is also below as it is experiencing growing pains. Smaller European countries are above the correlation, and this has often been the source of leveling accusations against the United States of "splurging" or wasting energy.

The reality, as usual, is a bit different. Comparison between United States energy consumption and that of European countries covering much smaller area, beyond "ball-park" statements, is not that meaningful when considering the fact that transportation accounts for about one-third of energy consumption. This point is made clearer in Figure 3.2, which graphs PPP GDP per capita vs. oil consumption per capita. Canada, for example, suddenly drops below the correlation, reflecting the country's vast area. It takes two hours to cross the Netherlands by car; it would require more than a week to drive across Canada.

Figures 3.1 and 3.2 can also be used to understand the concept of conservation. Much of the debate calls for the reduction of energy use (i.e., by moving from right to the left on the figures) but keeping the vertical axis constant. Because not even some of the most radical environmental organizations would advocate reducing the standard of living in countries such as the United States or in the European Union, conservation as a means to reduce total energy consumption is not supported by historical evidence.

This fact reflects the Jevons Paradox. William Stanley Jevons was a British economist who, in his classic *The Theory of Political Economy* (1871), suggested that as a society becomes more efficient, use of a commodity or a resource increases and then decreases. In the case of energy use this has been the obvious case for two reasons: 1) People the world over have understood the use of energy as integral to better living and 2) society constantly finds new uses of energy. While estimates vary, in the United States perhaps as much as 5 percent of power generation goes to computers, instrumentation and the Internet.

One hundred years before Jevons, the Scottish economist Adam Smith published the classic *The Wealth of Nations* (1776), identifying industrialization as the national trait that separates rich from poor countries. But in the twentieth century, industries have fled the rich countries – a process that has been ongoing for more than 50 years. The muscular allegories of "heavy" industries so widely espoused by communism have disappeared along with the regimes. The professed industrial prowess in the former communist countries has been discredited by images of shut down obsolete industries and horrific, unchecked, insidious pollution that filled the news media in the 1990s.

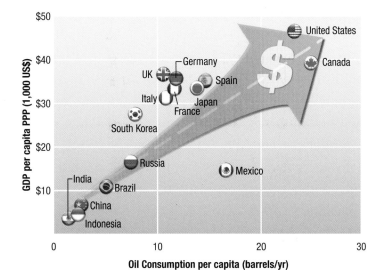

Figure 3.2. Gross domestic product per capita vs. oil consumption per capita (Source: EIA, CIA Fact Book, Time Almanac)

What is often missed in the debate is that the corollary from Figures 3.1 and 3.2 is also true. Use of energy generates wealth. At the beginning of the twenty-first century, energy consumption and access may have already replaced industrialization as the yardstick of the wealth and poverty of nations. The case for energy wealth and its significance is not difficult to make. Entire regions, such as the southern portion of the United States, owe much of their prosperity to the air conditioner, complete with its demanding energy use. Would Houston, Dallas, Atlanta and Miami – America's new cities – have reached their pre-eminence without heavy use of energy in the hot and humid climate conditions they inhabit? Not likely!

Energy Intensity
The Jevons paradox should not be confused with efficiency and, in fact, conservation. Improvements in efficiency in energy use have occurred over more than a century. The important concept in efficiency is energy intensity, which can

be defined as energy use per unit of GDP, e.g., BTU per dollar of GDP. Figure 3.3 is a plot of the energy intensity since the middle of the nineteenth century for the United Kingdom, the United States, Japan and the Developing World.

From the middle of the nineteenth century and the Industrial Revolution to 1880, energy intensity in the UK increased to a maximum; think of this as a "learning curve." Coal fueled the era. After 1880, energy intensity in the UK has largely declined, clearly implying that the nation has been becoming more efficient. Around World War I a significant "blip" was observed. That is when the then-Lord of the Admiralty had a fine idea: switching the British navy from coal-fired to diesel. Ships became far more maneuverable and faster, to run the proverbial circles around the German navy. (That man's name was Winston Churchill who went on to do bigger things eventually.)

The increase in energy intensity at the time was related to the learning curve associated with energy transition, a painful process always and a harbinger of change.

The United States took about 40 years longer than the UK to reach a peak in its energy intensity. The long "learning curve" is intimately connected with the nation's buildup, the "Go West Young Man," when poor European immigrants developed the vast new nation. Around the turn of the century, another pioneer had a fine idea. He thought that common people should have a car. Before then, vehicles were playthings of the aristocracy. By filling the need for an efficient way to traverse the expanses of the new country, this pioneer helped make America the mobile society that it still is today. His name was, of course, Henry Ford.

After the United States' energy intensity reached a peak, it started declining, through the Great Depression, through World War II and up to a plateau that lasted about 30 years, from the end of the War until the early to mid-1970s. That was when America splurged: when middle-class people moved from the cities to the suburbs, built large new houses and cars, and filled their lives with home appliances.

It is not clear whether the Arab Oil Embargo was the clarion call for efficiency or whether it was due to happen anyway, but starting from the mid 1970s and continuing to 2010, the US energy intensity has declined to less than half of what it was then. In other words, the portion of the economy that goes to energy has been reduced by more than half.

This is one of the reasons why gasoline retail prices, which escalated in 2007-2008 to the highest inflation-adjusted prices ever, failed to cause an economic upheaval or a substantial reduction in consumer use. Through the looking glass of the economy, gasoline at $4 per gallon would have had to rise to $7.50 to $8 to have the same bite as gas prices did in the 1970s.

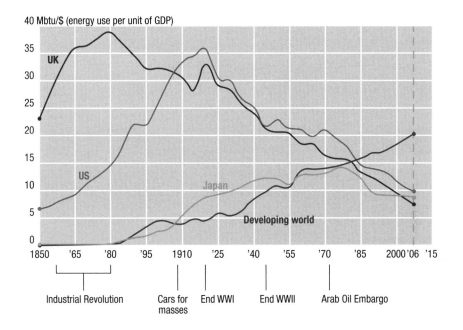

Figure 3.3. Energy intensity in the UK, USA, Japan and the developing countries (Source: DOE/EIA, Skov, NAS)

Japan reached a mini-peak in the early 1980s, after its meteoric economic growth propelled it into the highest levels of industrial and high-technology output in the world. But it has also started showing a decline in its energy intensity. It is very telling from Figure 3.3 that three major nations, the UK, the United States and Japan, with very different cultures, very different histories and, especially, very different geographies, all have roughly the same energy intensities today. If this is what people suggest as energy conservation, then these data offer a convincing proof. It is, of course, essential to note that all three countries have simultaneously experienced a continuous increase in total energy demand.

Finally, Figure 3.3 also should put to rest the popular and often-repeated refrain that the developed countries, especially the United States, waste energy. The developing world uses about twice the energy to generate a unit of their GDP than the developed countries. It is the developing world that needs to become more efficient.

The Big Players in Energy Consumption

Other than the energy-producing nations, which we have covered in Chapters 1 and 2, the world energy scene is dominated by the consumption of two countries: the United States, which has been the controlling market thus far, and China, which unquestionably will dominate in the future. It is worth comparing the countries' past, present and future energy mix, which appear in Figures 3.4 and 3.5.

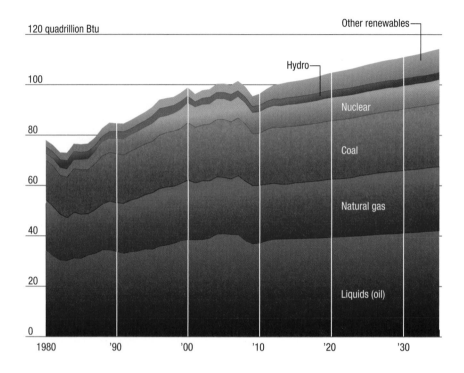

Figure 3.4. United States energy consumption and forecast
(Source: DOE/EIA, ExxonMobil)

Contrary to common misconception, the United States is not much different from the rest of the world. It uses more energy per capita, but the energy mix is very much like the world mix: about 40 percent oil and 23 percent each of gas and coal, for a total of more than 85 percent derived from fossil fuels. Again, by the year 2035, forecasts still suggest these fuels will overwhelmingly dominate.

China is an entirely different matter. Figure 3.5 shows an energy mix that is very different from the rest of the world. In China, coal accounts for an overwhelming 70 percent of the energy mix. However, oil is growing in some years at a rate that is 20 percent larger than the previous year; since 2000, the annual increase has almost always been by 10 percent.

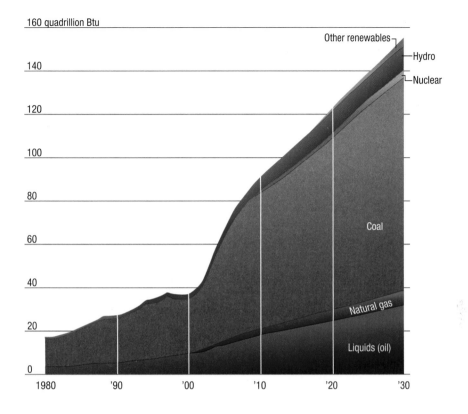

160 quadrillion Btu

Other renewables

Hydro

Nuclear

Coal

Natural gas

Liquids (oil)

Figure 3.5. China energy consumption and forecast
(Source: DOE/EIA, China Academy of Sciences)

Energy will be China's "choke point," and the search for adequate energy resources will be as important a priority for that nation as for any. And yet the Chinese, ever Confucian, suffer from three serious deficiencies:

First, the Chinese constantly publicize efforts in conservation and alternative energy sources. The reality is that the Chinese are adding, on average, more than one coal-fired commercial power plant per week.

Second, as we have mentioned earlier, conservation efforts have never achieved a reduction of total energy use. China has about 25 cars per 1,000 people, compared with the United States at more than 450 cars per 1,000 people.[6] Conservation certainly will not lead to a reduction in demand.

Third, China will not become a militaristic power anytime soon but will aggressively pursue energy resources throughout the world. Already, they have made investments from Canada to Australia to Argentina. More to the point, the Chinese will do business with regimes that are either hostile or unsavory to the West, such as Iran, Sudan and Myanmar.

Burdened by real environmental problems, some of which have reached catastrophic levels, China has become very defensive, occasionally making public statements that, if taken seriously, are nothing less than throwing the baby out with the bath water. "Pollution" thus becomes all-encompassing, covering everything from the real health issues of nitrogen dioxide, sulfur dioxide and particulates to the emissions of carbon dioxide.

Of course, nobody really believes that China will ever even come close to a carbon dioxide–constrained economy and energy use. In fact, the danger is that the almost certain failure to comply with any Kyoto, or successor, accords may muddle the issues and prevent real environmental stewardship in the form of cleaning up water bodies and rivers and, more urgently, improving air quality. China is in real need of this work, and it must act immediately and massively.

The problem is that over the last three decades, China's energy consumption has soared. But as is usually the case with developing countries (including the United States in its early development), energy use has been coupled with low energy efficiency.

This huge increase in energy use, along with abundant and cheap labor, allowed China to become a major exporter of industrial and consumer goods and has stimulated its economic development. Although the lives of the Chinese people have improved tremendously, the approach has also been associated with excessive energy utilization, energy shortages, and extreme environmental pollution. China's energy intensity (i.e., energy use per unit of the GDP) has been much higher than that of the US (see Figure 3.6). And while China's energy intensity is improving, cheap energy – subsidized and controlled by the government – has prolonged the country's inefficient use of energy and weakened the nation's development.

Furthermore, China's energy intensity has been increasing since 2001, an eminently undesirable event. Although China uses only about 75% as much energy as the US (about 75 quadrillion BTU vs. 100 quadrillion BTU in the US) the country's energy intensity puts China at a significant disadvantage: 13,800 vs. 8,800 BTU per dollar GDP. In other words, China uses about 60% more energy per unit of GDP than the US. It's also worth noting that on a per-capita basis, China's consumption is only about one-sixth that of the prevailing rate in the US.

Clearly, the Chinese economy has to be sharpened significantly to compete with developed nations, especially as imported energy prices will continue their upward trend and as increasing Chinese labor costs no longer offer major advantages.

After importing more and more energy from foreign sources at much higher prices than the controlled domestic ones, China can no longer afford to subsidize all energy components. Reforming energy pricing (which ultimately means raising energy prices) has become inevitable. In fact, some of the wealthier areas of the country, such as Shanghai, are already seeing higher prices. Of course, raising energy prices has long been the plan, but implementing it has not been easy. During good economic times, the government is afraid of market inflation; during the current economic downturn, personal and business financial sustainability is a big concern, which is very much tied to social stability.

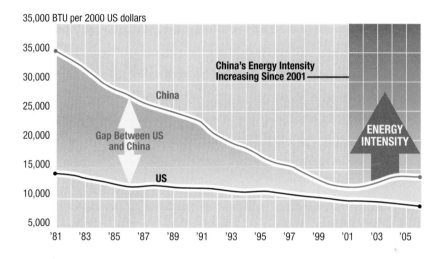

Figure 3.6. China energy intensity vs. the United States (Source: DOE/EIA)

Nevertheless, it is apparent that China's leaders have few options. Cheap energy appears to be a thing of the past for China. So, let's attempt to offer the Chinese a future energy blueprint:

First, recognize the intimate connection between energy abundance and the good life that the people desire. Resolving the certain, looming shortages should be the highest priority of the government, energy companies, and all national institutions.

Second, because of international competition and because of its clear aversion to confrontation, China should find answers inside the country. This implies that the plan to emulate the United States in all respects, a thinly disguised national obsession for three decades, will have to be radically readjusted. For example, China may be the first to truly electrify transportation, far ahead of the United States and Europe.

Third, if personal lifestyle and urbanization are to continue as unabated, worthwhile, and unstoppable processes, they have to be enabled by means other than importing ever-increasing volumes of oil. Coal, which the nation has in abundance, has to play an increased role, but nuclear certainly has to expand many-fold. Today, these energy sources contribute practically nothing to transportation. The only way to remedy this is by electrifying transportation or, in the case of coal, to embark on coal liquefaction and (even more bold) gasification, producing natural gas or hydrogen. Hydrogen could be used to power the vehicles of the future although it may not be the most attractive or practical option (See Chapter 9).

Such transition is painful anywhere, but in China it will have to be massive, on a scale never before attempted or fathomed. By 2020, as many as 40 percent of Chinese vehicles could run on electricity. This can be done either with batteries or by providing guideways where vehicles draw continuous electricity. The technology is already available, or nearly so. Because China's transportation sector is likely to grow at a rapid pace, building it using different infrastructure is only slightly more cumbersome than adding to existing infrastructure. China is starting at a far earlier point in its infrastructure development than the United States or Europe, which will work in its favor. Nations with fully developed infrastructure are burdened by it and are far more reluctant to change.

China can lead the world to its next stage by actually using its energy challenge to its advantage. It can turn a looming potential catastrophe into an opportunity.

Energy Interchangeability

 "We hold these truths to be self-evident, that all men are created equal," reads the beginning of the American Declaration of Independence. Unfortunately, for those wistfully seeking "Energy Independence," or more precisely "Fossil Fuel Independence," all potential fuels are not created equal. Fossil fuels have distinct advantages over alternatives in terms of energy potential.

One advantage is abundance. Coal is one of the most abundant energy resources, with supplies capable of meeting electricity needs for more than 250 years. And despite persistent claims by the media and peak oil theorists that we will quickly exhaust our oil reserves, the world is not likely to run out any time soon. Geopolitical factors, including production quotas among major oil exporting countries and supply bottlenecks in major importing countries, have fed recent price increases.

The abundance produces a second advantage – price. Ethanol, for example, continues to cost significantly more than gasoline on an energy equivalent basis, despite large tax credits for its production. Coal is still half as costly as nuclear for electricity generation. Wind and solar are too intermittent to serve as base-load power generators, and they cost two to five times as much per kilowatt hour of electricity generation when reliability is factored in.

Fossil fuels, in particular natural gas and oil, have enormous versatility, as Figure 3.7 illustrates. Byproducts from fossil fuels not only help defray their costs as fuels, they also make fossil fuels essential to a wide variety of industries. Only biofuels can provide similar byproducts.

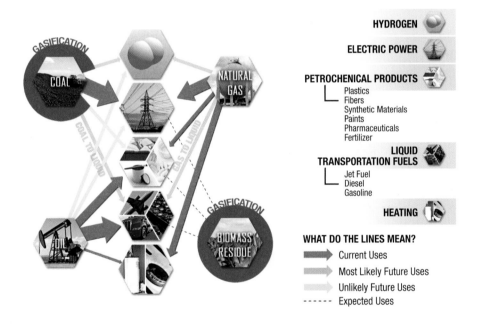

Figure 3.7. The versatility of fossil energy fuels (Source: Energy Tribune)

Fossil fuels such as oil, natural gas and coal can be used interchangeably, although with lesser levels of efficiency depending on the use. Coal is best used for electricity production, as it is cheapest. But it can be pulverized to produce a natural gas to heat homes. And through liquefaction it can, in a pinch, provide motor fuel for vehicles whose engines are modified to use it. Natural gas can be fired up to heat a home or a boiler for electricity. It can be reformed from a gas into a liquid through a well-known conversion process to fuel motor vehicles. Oil can be distilled into gasoline for transportation or burned directly to produce electricity (although virtually none is used to produce electricity since coal and natural gas cost less). Figure 3.8 shows the current use of fuels, and one of the most striking features is the lack of overlap of uses. The world uses practically no oil for power generation and uses essentially nothing but oil in transportation. This suggests that the use of alternative energy sources for power generation would save very little oil. Alternative fuels such as solar, wind, geothermal and nuclear can provide electricity, in most cases at considerably higher costs, but they cannot provide liquid fuels for transportation.

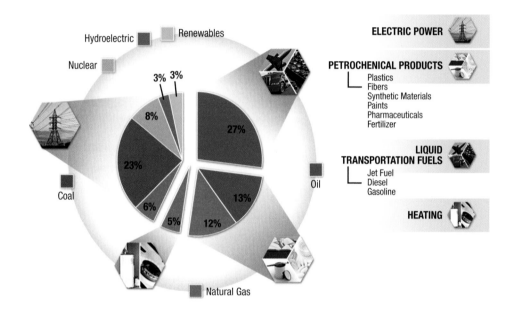

Figure 3.8. The lack of energy interchangeability (Source: Energy Tribune)

It is precisely for these reasons that many lawmakers and groups interested in achieving Energy Independence with reduced fossil fuel use are backing the use of so-called biofuels, such as ethanol, for transportation.

But in pursuing biofuels, proponents are ignoring many shortcomings that make them inadequate and even potentially economically and environmentally hazardous if used as anything more than a supplement to existing gasoline stocks. These alternative transportation fuels are discussed in more detail in Chapter 9.

Again, the sobering conclusion is that there are no alternatives to fossil fuels for decades. The world faces two challenges, and they are quite different. The first is to replace fossil fuels with alternatives. The second is to achieve more energy interchangeability: i.e., to use different fuels across consuming sectors. These issues will be explored in greater detail and our arguments supported as we delve deeper into the energy industry in subsequent chapters.

CHAPTER 4

Well Construction and Production

Chapter 4
Well Construction and Production

How are oil and gas accessed and produced? Early sources of oil and gas were identified by the presence of "seeps" at surface. While this may have proved to be a valuable phenomenon for early humans, it is certainly inadequate in today's quest for energy sources. The vast majority of oil and gas fields (at least, those of commercial interest) are far underground and owe their existence to the presence of sealing rock layers (called "cap rocks") that have acted as traps to accumulate the slowly migrating oil and gas beneath them. Leakage to surface from any one such deposit of hydrocarbon is relatively small. However, in aggregate, it is estimated that natural seeps release 600,000 tons of oil into the oceans every year.[7] Sensitive instruments can be used to detect volatile hydrocarbons in the atmosphere above potential oil and gas fields and have been used in aerial surveys. More recently, satellite imagery has provided information on oil and gas deposits from seeps visible from space.

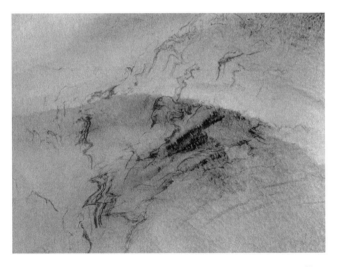

Figure 4.1. Natural marine oil seeps captured using satellite imagery

The most widely used technique to identify such promising prospects, is interpreting data from seismic surveying technology. Governments, oil companies and potential bidders for acreage all need this sort of information, and the seismic data collection and interpretation businesses are highly competitive. The technology involves the transmission of energy pulses (sonic energy) that propagate through the Earth, generating echoes that differ according to the physical properties of the rocks they encounter. Sensitive listening devices on surface (geophones) capture the echoes, and the data from many such devices are integrated to form a 3-dimensional "map" of the rock layers to depths of many thousands of feet underground. Seismic techniques can highlight geological features that are completely invisible at surface and, in particular, the tell-tale signs of potential traps or faults where oil and gas may be present.

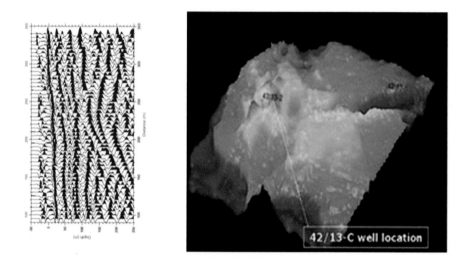

Figure 4.2. Two types of presentation of seismic data. On the left, shading provides information about rock properties and orientation. On the right, computer software analyses raw seismic data to produce a 3-dimensional representation of the subsurface structure.

If such hopeful signs are encountered, the next step is normally to drill an exploration well. This involves the use of a drilling rig, whose size and cost (or "day rate") will depend on a wide variety of factors, both technical and commercial. Very few oil companies own drilling rigs, although several governments (or their national oil companies) do. In general, though, the drilling rig is owned by another service company (the Drilling Contractor), and the oil company rents the rig equipment and crew on a daily basis, or on a long-term contract with an assured "day rate." Table 4.1 shows the kind of variations that can be seen at any given time for "day rates" of several types of the more frequently used offshore drilling rigs. By comparison, onshore (land) drilling rig day rates entering 2010 were typically between $20,000 and $25,000.

Table 4.1. Offshore Drilling Rig Day Rates (May 2010) (Source: Rigzone.com)

Rig type	Water depth *(ft)*	Day Rate *($)*
Drill ship	< 4,000	175,300
Drill Ship	> 4,000	423,252
Semi-submersible	1,500 +	301,684
Semi-submersible	4,000 +	411,207
Jackup IC	250	96,137
Jackup IC	300	111,966
Jackup IC	300 +	151,573
Drill Barge	< 150	30,000
Drill Barge	150 +	77,000
Inland Barge		49,921
Platform Rig		37,338
Submersible		35,500
Tender		119,773

Periodic day rates for any particular rig type also vary considerably. The main factor controlling the day rate is quite simply availability, which, in turn, is driven by the price of oil and gas. Rigs can be expensive to commission, and they take a long time to build, especially in the case of the larger offshore units. Drilling contractors are reluctant to build rigs without assured work since their arrival would only result in a softening of the market. However, building rigs only in response to tight market conditions almost inevitably leads to a shortfall in available units, due to long construction lead times. It may also lead to an oversupply of rigs, particularly if the industry enters a down cycle, by the time the rigs are built and fully commissioned. This, inevitably, leads to price collapse as drilling contractors struggle to keep their units working, to offset the cost of high initial investment and daily upkeep, regardless of absolute profitability.

The first commercial oil and gas wells were fairly simple affairs, drilled on land to only relatively shallow depths, a few tens to several hundreds of feet. These wells were usually no more than deep holes in the ground, with no major structural components. However, it quickly became apparent that, unless special precautions and procedures were undertaken, such wells would collapse or hydrocarbons would be lost by escaping from the well into other rock layers, contaminating ground water supplies or causing blowouts and "wild well" fires.

Figure 4.3. Deepwater drilling rigs – on the left a drillship; on the right a semi-submersible. Both can work in extreme water depths. (Source: Transocean, Shell)

As the economic, technical and safety benefits became more widely appreciated, wells began to be properly engineered with due regard for protecting ground water and preventing flow between rock layers (zones). This was accomplished by implementing techniques that remain largely unchanged today. With very few exceptions, oil and gas wells are drilled in several stages; each stage is complete when it has been "cased" and cemented. Wells are cased with sections (joints) of tubular steel pipe, threaded at each end so that they may be connected together (or made-up) to form a continuous steel conduit. This conduit (casing) is made-up, joint by joint, as the pipe is run into the well until its lower end reaches the bottom of the section that has just been drilled. At that point, cement is pumped down through the casing and out into the annular space between the casing and the drilled borehole. The cement fills the gap, displacing drilling mud ahead of it and then setting to form an almost impermeable seal around the casing. The next section of the well can now be drilled with a smaller drill bit, creating a smaller borehole which, in turn, is cased and cemented. This process is repeated until the well reaches its final target depth (TD). Thus, the traditional well consists of a series of "nested" steel casings, each one cemented inside the others and inside the borehole that it traverses.

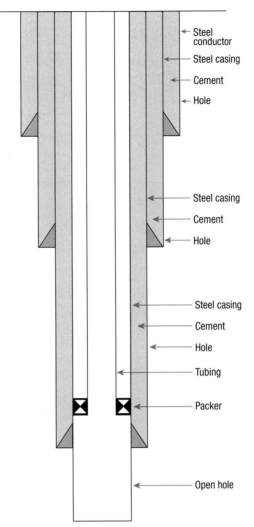

Ground Surface

Steel conductor

Steel casing

Cement

Hole

Steel casing

Cement

Hole

Steel casing

Cement

Hole

Tubing

Packer

Open hole

Figure 4.4. Schematic of a typical oil or gas well showing the arrangement of nested steel casings used to construct the well, isolating it and ensuring it maintains pressure integrity. These casings are cemented in place using high pressure cementing equipment.

The cement used to secure and seal these steel conduits in place is conventional Portland cement, of a type similar to that used in civil construction except that it has higher tolerance to certain salts found in sub-surface strata. Although the amounts of cement used in well construction are small relative to those used in civil engineering, the conditions of use (high temperature and pressure) require premium properties from the cement systems. Thus, the cement itself commands premium prices, and special additives are also incorporated in the designs to modify cement slurry behavior to suit specific well conditions. The final cement systems used may, thus, be quite complex and priced accordingly, making this a very attractive business segment for the companies providing this service.

More often than not, these days, the actual drilling process involves the use of a sophisticated array of mechanical and electronic devices attached to the bottom of the "drillstring," the name given to the steel tube, consisting of many joints of drill pipe, which connects the drilling assembly with the surface. The use of such hardware allows wells to be drilled in specified directions, often following very complex paths in three-dimensional space. Thus, it is possible to drill many tens of wells from a single offshore platform, for example, to intersect specific layers of a hydrocarbon reservoir, or to access several small reservoirs at varying depths and varying distances from the platform, without the wells colliding. Also, since the mid-1980s it has become routine to drill horizontally, creating extended-reach wells that either maximize contact between the well and the reservoir (thereby increasing production and recovery) or access hydrocarbons in difficult or sensitive environments. For example, some wells have been drilled from land to tap reservoirs that lie under the sea, and directional wells are routinely drilled from "concealed" locations to access oil deposits under cities like Los Angeles.

Several projects have demonstrated the ability to accurately drill wells with lengths in excess of 6 to 7 miles (10 to 11 kilometers) that intersect targets only 10 +/- feet (3+/- m) thick at depths in the range of 1-3 kilometers underground See Fig 4.5. Using variations of the same technology, it is also possible to drill multilateral wells, which, as their name suggests, consist of several individual drainholes emanating from a single "mother" wellbore at different locations and/or in different directions See Fig 4.6. Such wells can be costly to construct versus a normal well but they can offer much more production than a single well at lower cost than would be incurred drilling multiple wells from surface. They also reduce the impact of drilling on the environment by reducing waste, reducing material requirements and reducing operational footprint, while still accessing hydrocarbon reserves. Finally, and often most significantly, while the cost of such complex wells may be high, the overall cost of the project is likely to be much lower due to the simpler surface infrastructure required, compared with having multiple wells. Thus, drilling several complex wells from one facility will eliminate the cost of building the multiple platforms that would be required if only simple vertical wells were to be drilled to drain a large offshore field.

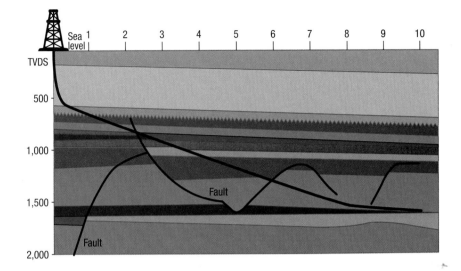

Figure 4.5. Depiction of an extended-reach well through multiple formation sequences. Extended-reach wells of this sort are used more and more to access oil and gas deposits in difficult locations.

Well depths are, generally, a function of how far underground the oil is located. Today, the typical range is 10,000 to 14,000 feet (~3,050 to 4,300 meters) although some exploratory wells have been drilled to depths approaching 35,000 ft (~10,670 m). As noted above, important technical advances have also been made in the drilling of so-called directional wells. The latter are not simply holes sunk into the Earth but rather are carefully designed conduits, tracked and steered in real-time with sophisticated sensors and telemetry packages, to intersect specific oil- and gas-bearing intervals, many thousands of feet under the ground.

As well depths have increased, temperatures and pressures have also risen, necessitating the use of new drilling fluids ("muds") to cool the drill bit and carry rock cuttings from the well, exotic cement systems, and casing and wellhead equipment made from special alloys. In the competitive world of the oilfield service businesses, all of these and many other challenges are rich grounds for creativity and innovation, where all players can strive to gain technical and commercial advantage. A novel tool or chemical additive, or a method to improve efficiency at the wellsite and reduce non-productive time (e.g., introducing a new, improved type of connection to join sections of casing together) can all reap substantial rewards for a company, increasing its market share, its profitability and its stock price.

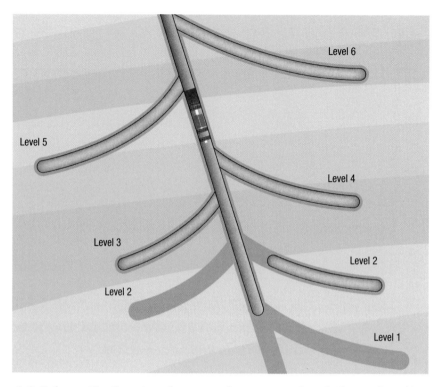

Figure 4.6. Schematic showing the general concept of multilateral wells and the 6-level nomenclature used to describe them. (Source: IDEAS)

If the drilling contractors and other service companies (which will be discussed later in this chapter) do their jobs properly, the well should reach the target, originally identified by seismic, without major incident.

The next step, then, is to establish whether oil or gas are actually present and, if so, whether they are there in significant enough quantities to be commercially viable. Identifying the presence of hydrocarbon is usually undertaken by service companies that specialize in a technique called "wireline logging." This involves lowering electronic devices ("tools") into the well on the end of an armored, multiconductor cable (the "wireline"). The tools measure a variety of physical characteristics, including electrical resistance of the rock layers, sound velocity through them, the presence of radioactive elements in the rocks, or the reaction of the rock to bombardment with particles from radioactive sources, usually neutrons. The tools transmit their measurements through the wireline to surface, where they are processed and analyzed. The wireline logging company can easily identify the presence of oil or gas and the exact position of such a "pay zone" in the well, in terms of depth and extent. The wireline log is used to correlate all the

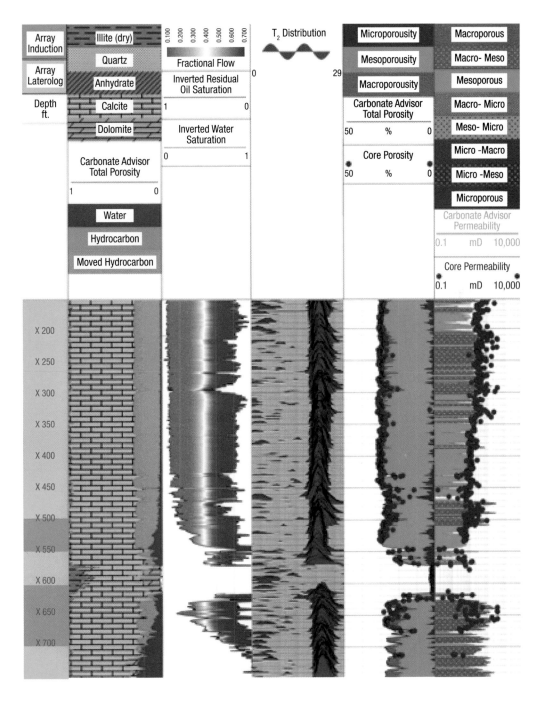

Figure 4.7. An example of a processed wireline log. Information from a variety of sensors lowered into the well on braided wireline is recorded and represented pictorially in this type of log, which can show measurements of permeability, porosity, oil, gas, water saturation, lithology, amongst other properties.

well depths and to confirm these for subsequent activities in the well, including perforating, the process by which the physical connection is made between the hydrocarbon-bearing rock and the inside of the cased, cemented well.

Although it is undoubtedly the most important component, the well itself is actually only a small piece of a vast and complex infrastructure required to exploit oil and gas. The well's function is to competently connect the oil or gas reservoir with the surface facilities, maintaining pressure integrity and preventing leaks to surface or between rock layers with different pore pressures.

From the wellhead, which is on the surface on land or on the platform or on the seabed offshore, the produced fluids are carried by steel pipeline to a processing facility where they are separated, mechanically, into their constituent parts (oil, gas, water, typically). Alternatively, for marginal producing wells, this separation may be done directly at the well site and the crude oil stored in a storage tank that is regularly emptied and the contents trucked to a refinery. In the case of offshore facilities, the processing may be done on a nearby central gathering platform so that loading of oil tankers may be done directly in the field. Alternatively, produced fluids can be shipped via pipeline to shore or directly to a refinery for processing into refined products.

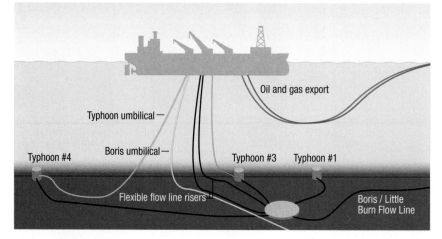

Figure 4.8. Offshore fields use a variety of development strategies. In this case, subsea wells are linked to a floating processing facility where oil and gas are separated and then shipped to shore via pipeline.

The flowlines (or pipelines), the gathering stations, compressor facilities and, in the case of offshore fields, the vast production platforms that separate, quantify and, to some extent, process gas and liquids, are all essential components of the production train. And, like other parts of the industry's infrastructure, they must be constantly upgraded and improved to deal with the challenges of a business that continues to push the technical frontiers, in the quest for new reserves.

Deciding on the type of platform to build in the offshore environment depends on many factors. Cost is obviously one of these, but there are others related to the environment in which the platform must operate. Water depth and seasonal weather patterns are clearly significant. Until recently, offshore production was restricted to continental shelf environments with water depths rarely exceeding about 650 feet (200 meters). In such scenarios, steel and/or concrete platforms supported at the sea bed were often effective and made economic sense.

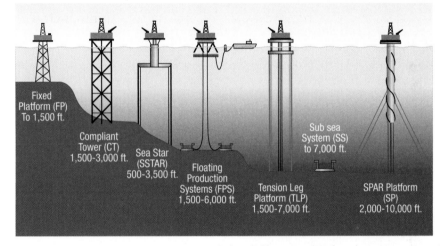

Figure 4.9. Schematic showing some of the different platform options available and the typical water depth for deployment of each type.

However, over the past 20 years, some of the most significant oil and gas discoveries have been made in much deeper water -- up to, or even exceeding, 10,000 feet (3,000 meters). In such a case , conventional platforms are impractical. For development and exploitation of such deposits, therefore, several alternative strategies have been employed to provide field infrastructure. Buoyant platforms tethered by steel cables to the seabed (Tension Leg Platforms), spars and Floating Production Storage and Offloading vessels (FPSO's) are amongst the solutions that have been applied successfully in deep water environments (See Figs. 4.10, 4.11, and 4.12).

An FPSO is, in essence an oil tanker fitted out with processing equipment to allow for onsite separation and storage of oil in remote or deepwater locations. They offer an elegant solution in several situations, including those where field reserves may not justify the cost of fixed infrastructure. Use of such vessels eliminates the cost to build a fixed processing platform or pipelines: When the field depletes, the FPSO simply moves to a new location. Oil is processed on-site and stored in the FPSO temporarily, until it can be collected and trans-shipped by a conventional oil tanker. Small fields can become commercially viable and larger fields in deepwater can be brought into production faster using the FPSO approach. As of this writing, the world's two largest FPSOs are the Kizomba A and the Agbami. Each has a storage capacity of 2.2 million barrels and in 2009 were operating in deep water off the coast of Angola and Nigeria, respectively. Each vessel is around 1,000 feet (300 meters) long and over 200 feet (63 meters) wide and cost close to $1 billion to build. Each can handle around 250,000 barrels of oil per day and can store around 2.2 million barrels of crude.

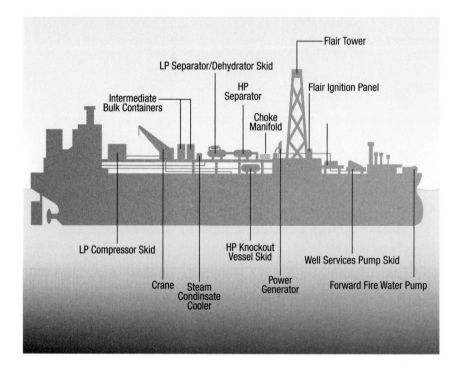

Figure 4.10. Schematic of FPSO

In so-called sub-sea developments, where wellheads are on the seabed rather than at surface, flowlines (in-field pipelines) are required to carry produced fluids from each wellhead to the production facility, regardless of whether an FPSO or a fixed platform is used. These flowlines often consist of special high-pressure hose, a composite of braided steel, carbon fiber and polymers, rather than steel pipe.

Export pipelines that transport the produced oil and gas to shore, on the other hand, use large diameter steel pipe, welded section by section in a continuous process, and then laid on the seafloor and covered with a protective layer of rubble. While offering an excellent means of transporting the hydrocarbons, pipelines are expensive to construct and commission. The steel cost alone is significant, and seabed mapping must be carried out to identify mountains, canyons and other obstacles that may lie along the projected pipeline route.

Figure 4.11. One of world's two largest FPSOs, Agbami, in deepwater off the Niger Delta, West Africa (Source: Chevron)

Producing oil and gas from offshore fields, in particular, requires significant support infrastructure. Manned offshore installations usually have at least one stand-by vessel, to provide for essential evacuation assistance in the event of catastrophe. Safety is paramount but, as demonstrated by events like the Piper Alpha disaster in the North Sea in 1988 and the Macondo well disaster in the Gulf of Mexico in 2010, even the close proximity of support vessels cannot eliminate the inherent risks of producing and processing large volumes of flammable hydrocarbons in a remote and hostile environment.

Apart from the essential stand-by boat, numerous supply vessels are required to ferry materials and personnel between the offshore installation and shore-based facilities. Such vessels transport spare parts, chemicals and equipment as well as fuel and provisions to and from the installation.

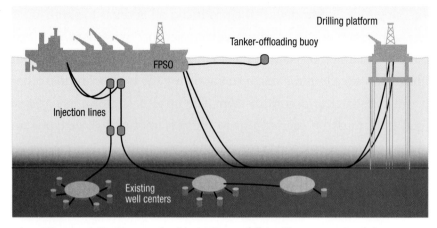

*Figure 4.12. Example depiction of flowlines required for a
subsea completion system*

As discussed, well architecture can be quite complex and wells are being drilled
to increasingly greater depths and with more creative geometries. To establish
a stable well system that can maintain its integrity beyond its useful life, many
individual steps are required: cementing the casing, strategic perforation of the
casing in exactly the right locations to establish effective oil and gas flow to the
wellbore (and avoid excessive production of unwanted water), and/or the placement
of screen "filter" systems that allow gas and/or oil to flow without filling the well
with sand – and all that is just to establish commercial production. To maintain
production over the long term, "artificial lift" is often required to help bring the
hydrocarbons to the surface. Even if a well flows naturally to the surface when it
is initially drilled and completed, it will do so for only a finite period of time before
the pressure depletes enough to require some artificial lift. Options include various
downhole pump installations or "gas lift" systems that help lighten and energize
fluids (oil and water) so they can rise to the surface.

Artificial lift is just one of the many specialized oil service businesses that
participate in the industry with the aforementioned drilling, cementing, wireline
and seismic services. The companies providing these additional services and related
technologies include those providing drilling fluids ("mud"), completion fluids, well
piping such as casing and production tubing (and the related handling services),
completion tools, perforating services, and the ever-increasing sub-sector known as
pressure pumping. The latter includes the exciting well stimulation service category,
providing a variety of creative mechanical and chemical well treatment applications
designed to enhance well production.

Figure 4.13. Onshore oil well pumping unit

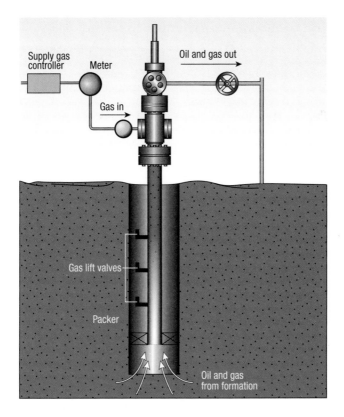

Figure 4.14. Gas lift schematic: injected gas through gas lift valves reduces fluid density (oil column weight – enabling flow to surface)

Including drilling and seismic services, there are perhaps a dozen sectors in the oilfield services industry. The services and technologies are diverse, but the significant players are mostly of US and European origin and base. Plus, for any particular sector, a limited number of entities dominate. Relative to major energy companies, the oil service sector often escapes public recognition. This is due, in large part, to the size of the largest service companies relative to the largest energy companies, as well as to the direct marketing and name recognition of the oil companies to the general public. Table 4.2 compares the most recent annual revenues of the companies comprising the American Stock Exchange Oil Index (OEX) and the Philadelphia Stock Exchange Oil Services Index (OSX).

Table 4.2. 2008 annual revenues: oil companies (OEX)
and oil service companies (OSX)

Oil Company (OEX)	Revenue ($billion)	Oil Service Company (OSX)	Revenue ($billion)
ExxonMobil	346.88	Schlumberger	27.16
BP	266.72	Halliburton	18.28
Total *	250.51	Transocean Limited	12.67
Chevron	184.13	Baker Hughes	11.86
ConocoPhillips	166.89	Smith International	10.77
Valero Energy	85.43	Weatherford International	9.60
Repsol YPF *	80.37	Nabors Industries	5.00
Marathon Oil	55.86	BJ Services	4.80
Sunoco	37.97	Noble Drilling	3.57
Hess	32.47	Rowan Companies	2.12
Occidental Petroleum	17.84	Tidewater	1.39
Anadarko Petroleum	12.48	Global Industries	1.07

* Total and Repsol YPF revenues estimated

Note the striking difference in gross revenues – by an order of magnitude – between the two major sectors. First on the oil service company list is Schlumberger, whose annual revenue would place it below ten of the twelve oil companies presently comprising the OEX.

Despite this disparity, the oilfield service industry is absolutely vital to the success and survival of the oil companies. The oil services sector is driven by oil company revenues and drilling activity; the health of the service companies is also linked to petroleum commodity prices, which naturally influence well exploration and development spending by the oil companies. The oilfield service firms are among the first to enjoy the benefits of increased spending – and are the first to suffer the effects of decreased spending. Oilfield services and equipment is thus a sharply cyclic industry.

The major integrated oil service companies (Schlumberger, Halliburton and Baker Hughes – the "Big Three") provide a broad range of oilfield services. Although they suffer along with their industry as a whole during downturns, their exposure is tempered by the greater array of services provided. The "boom or bust" cycles (especially "bust") can be more pronounced for other smaller "independent" service companies, as they are often regionally based or focused on providing one or more of a limited number of specialized services, or in a narrow application range. For example, the fortunes of a well stimulation company that provides hydraulic fracturing services primarily in US shale gas plays will rise and fall dramatically with natural gas price and localized rig counts. Thus, although these independent companies have historically posed a competitive threat to the major pumping service companies in localized areas, their success can also be fleeting.

However, emerging with strong built-in support and customer base are the so-called national service companies – those originating from within the national oil companies (NOCs) and those regionally linked to one NOC or another. As the global breadth of the NOCs increases, it stands to reason that their respective service arms pose a growing threat to the long-established, publicly held major service companies, especially the well stimulation providers that have been historically dominated by the few biggest players.

Well stimulation

As fields mature, especially outside of the operating worlds of the major producers and NOCs, and as the cost to produce each barrel of oil equivalent (BOE) increases, production enhancement through well stimulation grows in stature and importance. Most of the time, newly drilled wells require some stimulation to achieve production goals. And after a well is on production, it is unrealistic to expect problem-free production. First, wells are subject to natural production decline from reservoir pressure depletion over time as the well is produced. In addition, wells often suffer sharper (or accelerated) decline due to impairment of flow through the formation rock or the well itself. Sometimes these flow impairments stem from naturally occurring phenomena based on produced fluid properties, flow velocity, temperature and pressure changes; sometimes well operations induce plugs. Either way, the flow impairment is known as "formation damage." In addition, mechanical failures (tubing, casing collapses; pump failures, etc.) can cause stoppage of oil or gas production independent of formation damage and also require servicing.

Table 4.2. Example formation damage mechanisms for different well operations

Well Operation	Damage Mechanism
Drilling	- Solids invasion - Polymer plugging
Perforating	- Dirty perforating fluid - Compaction *(reduced permeability zone)*
Production	- Scale or salt deposition from water - Organic deposition *(wax, asphaltenes)* from oil - Fines migration *(sandstones)*
Workover	- Scale deposition from incompatible fluids - Formation of viscous oil-water emulsion
Stimulation	- Fines and sand production *(excess acid)* - Acid-formation reaction product precipitation - Scales *(acid and fracturing)* - Gel polymer plugging *(fracturing)* - Proppant fines *(fracturing)* - Viscous emulsion / acid sludge *(oil wells)*

No well operation can be considered truly non-damaging. Any invasive operation may damage well productivity. Production itself can cause formation damage; problems in oil and gas wells are often related to the onset of water production. As wells mature, water production typically rises as accompanying oil or gas production declines, and can become a serious, if not debilitating, issue. Water production can simply overwhelm production of the valuable hydrocarbons, or it can impair hydrocarbon flow through a variety of mechanisms: release of and plugging by formation fines (very small particles of rock found in sandstones); increase of water saturation (portion of pore spaces containing water) that decreases relative permeability to oil or gas; deposition of scales in the well tubing, pumps, and possibly in the formation; or deposition of salt from highly saline formation water as it travels up the wellbore and cools below its "salt saturation" limit at reservoir temperature.

Figure 4.15. Wax (paraffin) deposit on pump rod in oil well – can be removed with hydrocarbon solvents (Source: Production Enhancement with Acid Stimulation, Kalfayan, 2007)

Figure 4.16. Asphaltene deposit recovered from well. This kind of damage can be removed with aromatic or terpene-based solvents (Source: Production Enhancement with Acid Stimulation, Kalfayan, 2007)

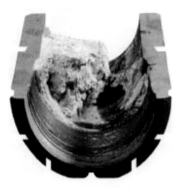

Figure 4.17. Calcium carbonate scale buildup in section of well casing can be removed with hydrochloric (HCl) acid or chelating agents (from Clariant)

Figure 4.18. Sand proppant pack containing fines (from proppant crushing – restricting flow. Such fines can be at least partially removed with mild acid solutions containing hydrofluoric (HF) acid (Source: Production Enhancement with Acid Stimulation, Kalfayan, 2007).

Well stimulation applications include removing (or dissolving) formation damage, bypassing damage (through fracturing or creating channels through the damage), and stimulating the oil- or gas-bearing rock formation beyond its natural (undamaged) flow capacity. These well treatment services are provided by the pressure pumping service companies, including those smaller entities focused solely on acid stimulation, fracturing, or both. There are three general categories of well stimulation (or well chemical treatments): remedial stimulation, matrix stimulation and fracture stimulation.

Remedial stimulation involves removing flow restriction in the wellbore or formation in order to re-establish undamaged production rate. Typically, operations are performed with acid or non-acid solvents and additives.

Examples:
- Aromatic solvent to remove wax *(paraffin)* or asphaltene deposits
- Solvent *(water or hydrocarbon)* plus additives *(alcohols, surfactants)* to remove "water blocks" (retained water in formation pore spaces)
- Acid *(e.g., hydrochloric)* treatment to remove scales; fines; rust and debris; and gel polymer damage caused by drilling mud, fracturing fluids, etc.
- Inhibitor chemicals injected deep into the formation to prevent scale or organic deposits
- Chemical treatment to reduce unwanted water production *(water shut-off or water control treatments)*

Matrix stimulation treatments most commonly use acids and are pumped at pressures that allow the fluid to penetrate but not fracture the rock formation.

Examples:
- Sandstone acid stimulation with hydrofluoric (HF) acid to dissolve minerals (and fines) found in sandstones (e.g., quartz, clays, feldspars)
- Carbonate acid stimulation with HCl or organic (e.g., acetic, formic) acid blends or non-acid chemicals that dissolve carbonate mineral to create extended channels from the wellbore into the formation

The primary remedial treatment and matrix stimulation method is "acidizing." The discovery of acid stimulation in 1895 – credited to Herman Frasch, a Standard Oil chemist who patented the use of concentrated hydrochloric acid to stimulate brine wells in Ohio – started a path to the acidizing boom of the early 1930s, initiated by a collaboration of Dow Chemical and the Pure Oil Company in stimulating oil wells producing from a carbonate formation in Michigan. The remarkable success of these treatments led to an overnight boom and birth of a new industry – well stimulation. Formed were the Dow Well Services company (eventually Dowell – later to be absorbed in Schlumberger), the Process Engineering Company (a predecessor of BJ Services, now part of Baker Hughes), and the stimulation division of the Halliburton Cement Company. These were among the only survivors, ultimately. Many other stimulation companies have come and gone in this highly competitive business sector, some disappearing within just a few years. The acid stimulation industry branched out to include the use of hydrofluoric (HF) acid in sandstones in the late 1930s, finding its success in removing drilling mud damage. The early acid treatments led to the observation of unintentional formation fracturing, which led to the development of hydraulic fracturing in the late 1940s.

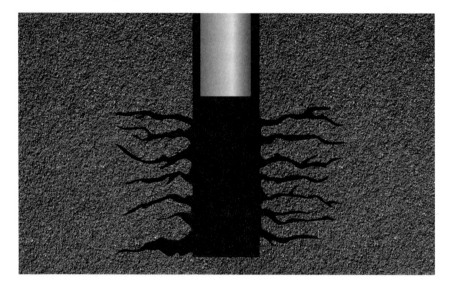

Figure 4.19. Carbonate matrix acidizing: acid exits perforations and creates channels ("wormholes"), thus establishing conductive flow paths from the formation to the wellbore

Today, acidizing technologies and methods range from carefully designed and strategically injected hydrofluoric acid treatments to remove fines from sandstone formations (without otherwise altering the rock properties) to large matrix acid treatments in carbonates to create a network of branched "wormhole" channels extending into the formation to the novel use of coiled tubing (long, unjointed pipe) and special nozzle tools through which acid is injected to create strategically placed tunnels extending tens of feet away from the wellbore in wells completed openhole (no casing) in carbonate formations.

Fracture stimulation, or hydraulic fracturing, refers to the injection of fluid (usually water-based or occasionally oil-based) into a well at rates sufficient to fracture the rock formation and to enable continued fracture growth (propagation).

Included at some point in the injection process and blended into the fracturing fluid is "proppant," a solid material such as sand (including resin-coated sand), bauxite, ceramics, resin-coated walnut hulls, and others. Sand is the most common and least expensive proppant. Deep wells that are under greater pressure require higher-strength proppants such as bauxite and certain ceramics. The purpose of the "proppant" is to fill the created fracture and "prop" it open, preventing it from closing after the stimulation treatment is complete and pressure is released. What remains, ideally, is a permanently open, highly conductive fracture (or multi-fracture) channel from deep in the formation to the wellbore, through which oil or gas can produce at commercial rates.

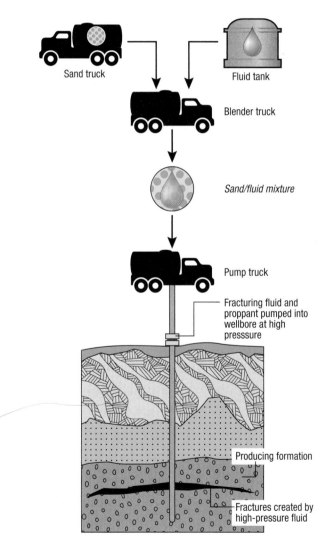

Figure 4.20. Hydraulic fracturing depiction (Source: FracMaster).

A sub-category of fracturing is "acid fracturing" in which the fracturing fluid (or at least one of multiple fracturing fluids injected) contains acid, which creates highly conductive "acid-etched" fracture channels. Acid fracturing is predominantly applicable in carbonate formations, which have high solubility in acid such as hydrochloric (HCl). Acid fracturing treatments do not require proppant, although proppant may be included depending on the anticipated stability of the acid fracture created.

On one hand, modern fracturing treatments are brute-force, high-horsepower, "frac jobs." On the other hand, they are expertly designed and engineered with the aid of complex mathematical computer simulation and prediction models; with careful specification of fracturing fluid and proppant properties, multiple treatment injection stages, volumes, and injection rates in order to achieve the stimulation objectives of fracture length, height, conductivity, and ultimately, well response.

From the first intentional frac job of an oil well in 1947 – in the Hugoton gas field in Kansas – the technology of hydraulic fracturing has come a long way. Today it is the single most prolific well stimulation technique, arguably applicable in all wells if mechanically and logistically possible.

Today, fracturing treatments can cost from a few tens of thousands of dollars to millions of dollars. The range of applications begins with small skin-bypass jobs that penetrate only a few feet beyond the wellbore to bypass localized formation damage and extends to modestly sized (average 50 feet in length) offshore "frac packs" for the stimulation and control of sand and fines production in high-permeability, poorly consolidated sandstone formations and large, viscous "gel fracs" carrying specialty proppants. At the upper end of the fracture application range are the massive, multi-stage, horizontal well fracturing in "tight gas sand" formations and shales, which include the popular "slickwater fracs" that pump massive volumes of water (sometimes millions of gallons) containing only a friction reducer additive (polymer) and proppant (e.g., sand) at maximum injection rates to create complex fracture networks throughout the formation, sometimes in more than one well at a time ("Simulfrac").

Fracturing is gaining particular attention today because of its widespread and critical application in achieving economic production from gas shales. Gas shales are alumino-silicate rocks containing clay minerals and particles of quartz (silica), feldspar, calcite (calcium carbonate), dolomite (calcium-magnesium carbonate), and other minerals. Shales that hold natural gas (or more rarely, oil) are very low in permeability, but they are rigid and brittle and thus amenable to hydraulic fracturing and subsequent retention of open fractures after stimulation. They are also prolific natural gas producers. As discussed in Chapter 2, shale gas has significant future energy implications. This makes shale gas fracturing one of the most exciting niches of pressure pumping services and represents the most recent boom industry within the oil services sector.

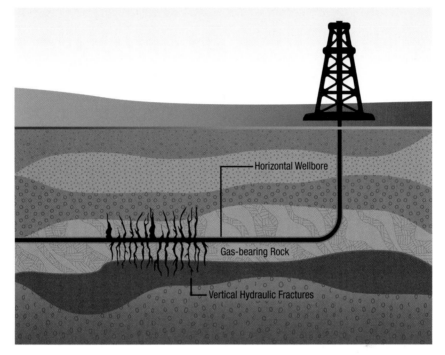

Figure 4.21. Depiction of horizontal well in a shale zone – with multiple vertical hydraulic fractures along the horizontal wellbore (Source: Energy Tribune).

Technology

Well drilling, completion, and production is becoming more costly as new frontiers and developments increasingly take place in extreme environments (deeper wells, in deeper waters offshore, higher reservoir temperatures and pressures, remote locations, cold weather conditions, etc.). With that, conventional wells and stimulation techniques find themselves less applicable. As the relative number of complex deepwater installations, long multilateral wells, and so-called "smart," self-maintaining wells trends upward, conventional well stimulation services – traditionally a staple of the major oil service companies – must be replaced or substantially and creatively modified. The challenge to the oil services industry is to provide production enhancement services that do not depend on conventional pumping equipment, chemicals, and well intervention methods.

CHAPTER 5
Bulk Transportation and Facilities

Chapter 5
Bulk Transportation and Facilities

One of the reasons that oil became the pre-eminent fuel of the twentieth century was simply that it exists as a liquid. This makes it relatively easy to extract from deep underground and to transport to a final destination. It can be pumped for long distances in pipelines or can be carried across vast distances in all manner of ships, trains, barges or trucks. Even more important, its refined products are mostly liquid, relatively stable across a wide temperature range and can be dispensed easily into vehicle fuel tanks, with relatively simple equipment and common sense precautions, at gasoline stations, airports, truck stops, etc. anywhere in the world.

Figure 5.1. Convenient, simple and safe – filling up the tank with liquid fuel.

Coal on the other hand, being a solid, must be physically mined and generally must be shipped in its natural state. It cannot be pumped, unless it is crushed and slurried using special techniques, and its mineral content produces significant amounts of ash, making it generally unsuitable for use as a fuel in vehicles. Such handling issues were, in fact, amongst the main reasons for the move from coal-fired external combustion engines to oil-fired external and internal combustion engines in trains and ships.

*Figure 5.2. Coal is cheap and plentiful but solid and dirty with
lots of ash to clean up.*

Natural gas, in some ways, is even more logistically challenging than coal. It is
volatile and easy to extract from underground, where it often occurs along with, or
dissolved in, crude oil and from which it can easily be separated. It can also occur
alone, as "dry gas" or associated with other volatile, low-molecular-weight liquid
hydrocarbons (condensate). Natural gas is composed of a mixture of primarily
methane with small variable amounts of ethane, propane and butane and, while
the latter two can exist as liquids at room temperature under modest pressures,
methane cannot. Thus, gas is much more difficult to transport and handle than oil.
It must either be shipped directly from the field by pipeline or must be liquefied or
compressed. Several techniques for transporting gas are in use today, and some new
methods are beginning to emerge, to compete with the established techniques.

However, even if infrastructure exists to move gas from the supply point
to the consumer, its end uses are sometimes limited without special facilities or
modifications to existing equipment. Thus, gas, while an ideal and cleaner fuel for
engines, requires special pressurized equipment to dispense it and contain it. Also,
because it has a lower energy density, volume for volume, fuel tanks must be larger
to accommodate the reduced distance attainable on a specific volume of fuel.

Despite these drawbacks, natural gas is poised to increase its share of world
energy provision very significantly over the next 10 to 15 years. This is partly driven
by the fact that gas is a "cleaner" fuel than coal or oil. It has a lower carbon-to-

hydrogen ratio and, therefore, when it burns it produces less carbon dioxide than the other fuels. This is seen as a good strategy to reduce greenhouse gas emissions. Also, gas is extremely abundant compared to oil, and vast conventional reserves are present in various countries, particularly Russia, Qatar, Iran and Canada. New technology has also provided access to huge gas reserves in unconventional low-permeability reservoirs where technology has allowed commercial extraction that would have been previously uneconomic. Finally, the establishment of a viable trading market for Liquefied Natural Gas (LNG), significant increases in supply (in terms of LNG production "trains"), and an ever increasing fleet of LNG bulk carriers have made it feasible to transport hydrocarbon gas from one continent to another without building thousands of miles of additional pipelines.

Figure 5.3. Vehicles converted to run on natural gas include large, pressurized fuel tanks.

Regardless of the reasons, such an increase in gas demand will still require enormous investments in transportation and distribution infrastructure, as well as end-user equipment. Japan, which was largely responsible for the development of the commercial LNG business as a cheaper alternative to high-priced oil in the 1970s, due to its own shortage of hydrocarbons, already has extensive infrastructure in place, as does Korea. But the countries of Western Europe, the

United States and China, for the most part, do not. Additional LNG carriers, storage facilities, re-gasification plants to convert the LNG back into gas that can be fed to a conventional distribution grid, will all be required.

This chapter examines traditional fuel transportation and distribution infrastructure and how it has evolved over the years. It also identifies future growth areas where investment is needed if the industry is to satisfy demand. Some current and emerging technologies that might contribute to future commercial exploitation of otherwise stranded resources are also highlighted.

Historical

At the beginning of the Industrial Revolution, factories were established in places where energy supplies were readily available. Fast-flowing rivers and locally-mined coal in the English Midlands and southern Scotland provided the bulk of the energy to turn water-wheels and run factory steam engines for sawmills and looms. Coal was also used to fire furnaces turning local metal ores into lead, copper, cast iron, steel and other important commodities. Similar scenarios evolved over the ensuing decades in Wallonia (Southern Belgium) and in Germany's Ruhr Valley as industrialization gradually spread across continental Europe. Initially, coal also fired steam-driven locomotives and ships to provide mass transportation. However, with the increasing availability of cheap liquid fuels derived from a rapidly expanding oil business in the early twentieth century and the simultaneous development of the internal combustion engine, gasoline and diesel quickly supplanted coal for many transportation applications. Initially, shipping and storing flammable liquid fuels was more troublesome than was the case for solid, stable and familiar coal, but the advantages, both from a technical perspective and in terms of convenience, far outweighed these minor considerations, particularly for smaller vehicles.

Pipelines

In the early days of the industry, oil was stored and shipped in wooden barrels and hauled, by teams of horses and wagon, the relatively short distances to loading points for river or marine transport further afield. The traditional haulage entrepreneurs, called "teamsters," a name that is still used for truckers and the labor union that originally represented them, were a critical and unpredictable element in the supply chain from wellhead to railhead or shipping port. Their control of this overland shipping market allowed them to charge oil producers sometimes extortionate prices. With oil prices fluctuating wildly between 10 cents and 12 dollars per barrel, oil producers were at their mercy. In 1865, in response to high shipping prices, Samuel van Syckel, an oil trader, constructed the first pipeline to carry oil the 5 miles (8 kilometers) from the Pithole field in Pennsylvania to a railhead loading point at Miller Farm. Despite sabotage and other difficulties, the pipeline proved its worth, routinely transporting around 2,000 barrels of oil per day at a tariff of $1/bbl. A second pipeline was duly constructed within two months, effectively signaling the end of the teamsters' monopoly. Following this success, pipelines quickly became a standard feature in transporting oil across relatively short distances of a few tens of miles. Oil pipelines of hundreds, and thousands of miles in length have since been built in and across many countries around the world. Van Syckel's original oil pipeline was only 2 inches (~5 centimeters) in diameter and made of wrought iron. Modern carbon steel oil pipelines can range from 4 to 48 inches (~10 to 120 centimeters) in diameter. They transport oil to consumers from giant oilfields offshore in places like the North Sea and Gulf of Mexico, from remote desert and jungle locations in places like Saudi Arabia and Indonesia, respectively, and from arctic fields on Alaska's North Slope.

Figure 5.4. Pipelines are sometimes laid on surface in remote or uninhabited areas The Trans-Alaska pipeline, pictured here, was raised above ground level to avoid melting of arctic permafrost. Most pipelines are buried, however, and are hardly noticeable.

Similarly, and of no less significance, an ever-increasing network of high-pressure gas pipelines carry natural gas from producers to consumers. Initially, gas pipelines were local affairs or, at the limit, were confined within single countries. In recent years, however, an extensive international grid of gas pipelines has been constructed, sometimes carrying gas thousands of miles from producing countries to consumers, across multiple international borders. These gas pipelines range in diameter from 2 to 60 inches (5 to 150 centimeters), and some operate at pressures of up to 1500 psi (~10 MPa). Thus, huge volumes of gas can be safely transported by pipeline to the USA from Canada and to Europe from Algeria (via Spain), Libya (via Italy) and from Russia (via Belarus and Ukraine).

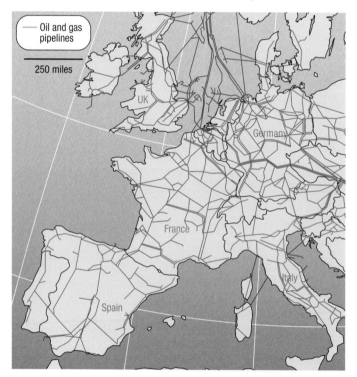

Figure 5.5. Europe has an extensive network of oil and gas pipelines and due to its dependence on foreign sources of natural gas, it currently has an annual pipeline import capacity of around 360 billion cubic meters, about 60% of which comes from Russia. Other major suppliers to the European pipeline network include Norway and Algeria.

Pipeline-delivered gas has largely dominated international gas trade, accounting for some 75% of all international flows. Historically, the bulk (about 70%) of this has been transported by international pipelines in North America and in Europe. In Europe alone, an extensive interconnected network stretches over more than a million kilometers (more than 625,000 miles), linking Russia to Ireland and from Norway to Spain.

In late 2009, China, too, began importing gas through one of two newly constructed 42-inch pipelines that run from Turkmenistan (via Uzbekistan and Kazakhstan). The estimated annual capacity for these lines is some 40 billion cubic meters or 1.4 trillion cubic feet of natural gas per year, the majority (75%) from Turkmenistan, the balance from Kazakhstan.

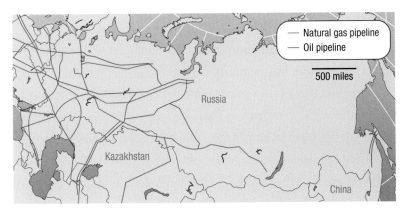

Figure 5.6. Russia's most important gas and oil pipelines

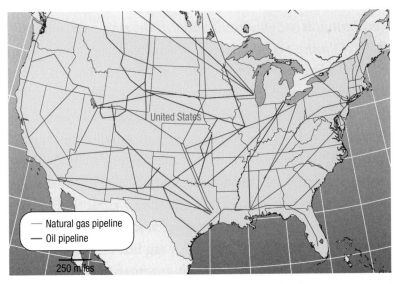

Figure 5.7. The US has a highly developed and extensively interconnected oil and gas pipeline grid with major imports of gas, primarily coming from Canada and Mexico. In the future, LNG will provide additional import capacity and distribution will be via the existing pipeline network.

Such pipelines have an enormous economic impact for both the producing and the consuming countries. For the producers, they provide valuable income, allowing the trade and export of a valuable commodity that might otherwise be flared or left unexploited. For the consumers, they provide a cleaner energy supply to heat homes or produce electricity, allowing dirtier, coal-burning power stations to be decommissioned.

Tankers

The most familiar transportation method for crude oil and other liquid petroleum products is the marine oil tanker. These ships transport the vast bulk of the world's crude oil from producing countries to those that consume it, as well as transporting refined products (gasoline, kerosene, naphtha, etc.).

The first truly successful oil tanker was built in 1878. As noted elsewhere, oil was a relatively specialist product in the nineteenth century and was shipped in wooden barrels along with other cargo. The barrels themselves were expensive, heavy and difficult to handle (even when empty) and sometimes leaked. The high cost and inconvenience of this method gave birth to the idea of a dedicated vessel to ship oil products in bulk. Ludwig Nobel (brother of Alfred, of Nobel Prize fame) designed the ship, the *Zoroaster*, to carry kerosene in specially designed interconnected iron tanks. The *Zoroaster* mainly plied the Caspian Sea around Baku but was capable of sailing from Sweden (where she was built) via the Baltic Sea and the river Volga. While usable, however, Nobel's vessel had stability problems, due to sloshing of the bulk liquid as the ship moved (a phenomenon known as the *free surface effect*). This problem was solved by Colonel Henry Swan, working for Nobel, in 1883 when he designed a series of single-hulled oil tankers. His approach separated the cargo into smaller tanks, subdivided by transverse and longitudinal bulkheads to minimize unconstrained movement of the cargo. The technique is still used to this day.

Thanks to such innovations, it was possible to build bigger and bigger tankers to keep pace with the ever-increasing demand for hydrocarbon fuels as the oil industry itself evolved and expanded. Indeed, the ability to design and build reliable tankers helped in the establishment and development of one of the largest and most successful oil companies. During a trip to the Black Sea in 1890, Marcus Samuel, a young British entrepreneur whose father had established a successful business trading with the Far East, recognized the potential of shipping Russian oil to the Orient through the Suez Canal. The canal authority had rejected earlier attempts by companies like Standard Oil to move oil through the canal as too risky. Samuel's bright idea was to ask the Suez Canal Company to define specifications for the type of ship it would grant passage to, which it duly did. Armed with these, Samuel commissioned the building of three dedicated tankers. The first of these, the *Murex*, became the first oil tanker to pass through the canal with bulk Russian kerosene, bound for Bangkok and Singapore, in 1892. The company set up oil storage facilities in Asia and, initially, called itself The Tank Syndicate. However, by 1897, Samuel had renamed the company Shell Transport and Trading. In 1903, he joined forces with a competitor, Royal Dutch Petroleum, which was successfully producing oil in Sumatra. The companies merged formally in 1907 and the pecten shell (scallop) logo became one of the most recognized corporate brands in history.

Figure 5.8. Ludwig Nobel's Zoroaster, the first oil tanker in the world, had a capacity of only 240 tons of kerosene.

Nobel's *Zoroaster* was capable of transporting a mere 244 tons of kerosene (~1,850 barrels) while Samuel's *Murex* could carry more than 10 times as much: over 3,000 tons. Tanker size and capacity continued to increase as the international trade in oil flourished in the first part of the twentieth century. Until 1956, the largest tankers were generally vessels designed to navigate the Suez Canal and, indeed, two-thirds of Europe's oil passed through the canal and accounted for over half of its traffic. However, politics, rising anti-colonialism, growing Arab nationalism due to the creation of the state of Israel, Cold War rivalries and, finally, the withdrawal of promised funding by the US and Britain for Egypt's proposed High Aswan Dam all played a role in the surprise nationalization of the canal by Egypt's president Gamal Abdel Nasser in 1956. The subsequent ill-advised military action by an alliance of British, French and Israeli forces in a devious plan to retake control of the canal was disastrous. The Egyptians deliberately sank over 50 ships along the canal and used cranes and other paraphernalia to complete its blockage, resulting in closure of the canal until 1957. As a consequence, oil tankers had no option but to take the much longer route around Africa via the Cape of Good Hope. This, in turn, led to the commissioning of much larger vessels to carry crude oil in an effort to reduce cost.

Figure 5.9. Marcus Samuel, founder of the Shell Transport and Trading Company, later to become Royal Dutch Shell, one of the largest vertically integrated energy companies in the world.

Tankers steadily increased in size from the 1950s, leading to the construction of so called "supertankers," an unofficial and rather fluid term to describe ever larger vessels, and culminating in the building of Very Large Crude Carriers (VLCCs) and Ultra Large Crude Carriers (ULCCs). Such vessels include the largest ships ever built and are capable of carrying 1 to 3 million barrels of crude oil (~160,000 to 500,000 cubic meters). Their enormous size precludes the use of such ships in a variety of situations. For example, they are unable to dock at many ports, and may be required to offload cargo onto smaller vessels (lightering). They are also generally unable to pass through restricted waterways. The largest vessel of all, the *Seawise Giant*, built in 1979, was too large to pass through even the English Channel.

Figure 5.10. Gamal Abdel Nasser, the Egyptian President, seized control of and nationalized the Suez Canal in 1956 leading to the Suez Crisis, a military fiasco by British, French and Israeli forces that resulted in the canal's closure and forced tankers to sail around Africa.

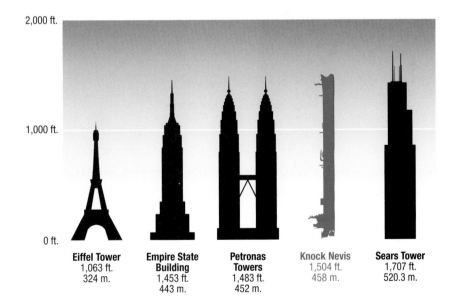

Figure 5.11. Size comparison of famous skyscrapers with the supertanker Knock Nevis, or T.T. Jahre Viking, the largest ship ever built . The length of this supertanker in the ULCC class is 1504 ft (458 m).

Efforts to standardize tanker (and other freight vessel) fleets and categorize them on their ability to negotiate the most common restricted waterways led to classifications like Suezmax, Panamax, Seawaymax and Malaccamax. These terms specify the naval architecture of vessels (in terms of such characteristics as length, beam (width), deep draft (distance from water surface to deepest part of submerged hull) and air draft (distance from water surface to highest point of a vessel) that will allow them to pass through, respectively, the Suez Canal, Panama Canal, St Lawrence Seaway (Great Lakes) and Straits of Malacca.

Table 5.1 Specifications of Petroleum Tankers

Petroleum Tankers					
Class	Length	Beam	Draft	Typical Min DWT	Typical Max DWT
Seawaymax	226 m	24 m	7.92 m	10,000 DWT	60,000 DWT
Panamax	294.1 m	32.3 m	12 m	60,000 DWT	80,000 DWT
Aframax				80,000 DWT	120,000 DWT
Suezmax			16 m	120,000 DWT	200,000 DWT
VLCC (*Malaccamax*)	470 m	60 m	20 m	200,000 DWT	315,000 DWT
ULCC				320,000 DWT	550,000 DWT

The smaller vessels are used for coastal transport, usually of refined products while the largest vessels are strictly ocean-going and may be unable to berth at some ports.

Estimates of the size of the world's total petroleum tanker fleet vary but, including coastal vessels hauling refined products, it numbers about 4300 vessels.[8] Many older vessels, including the largest ULCC ever built, the Seawise Giant, have been retired and scrapped or converted into floating storage vessels in deepwater oilfields. One of the reasons for the retirement of these older vessels, despite the growing world demand and consumption of oil, is the enactment of new laws in many countries requiring the use of so-called "double hull" vessels to provide added protection in the event of collision or other marine disaster. In older, single hulled vessels, the ship's hull acts as part of the oil storage tank. Thus, if the hull is penetrated by collision, oil leaks from the vessel directly into the sea, causing serious pollution. In double hulled vessels, by contrast, the oil storage tanks are discrete from the hull itself so that, in the event of a breach, oil need not necessarily spill from the vessel. It is not unanimously accepted that double hulled vessels are truly safer than their predecessors but the general consensus is that they do provide an extra level of protection and an increasing number of countries require their use today.

Over the years, there have been several major oil tanker disasters involving collisions or extreme weather events. The *Torrey Canyon*, the *Amoco Cadiz*, the *Exxon Valdez*, the *Braer* and the *Prestige* are among the names that may be familiar to readers. Of these, one of the largest ever tanker incidents involved the *Amoco Cadiz*, which ran into trouble in heavy seas off the French coast in 1978 when her rudder was damaged. The ship drifted out of control for several hours, eventually running aground. She was carrying almost 220,000 tons (~1.6 million barrels) of crude oil and the entire cargo spilled when the ship broke up on the rocks. The slick created was driven onto the French coast by strong winds and heavy seas and contaminated 200 miles (320 kilometers) of coastline, including valuable commercial shellfish beds and tourist beaches.

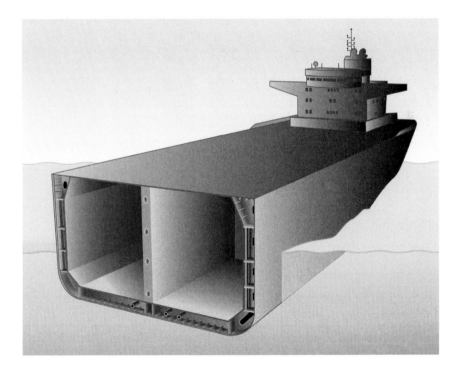

Figure 5.12. Basic configuration of a double hull tanker. The oil storage tanks are separate from the hull of the ship reducing the risk of damage to the tanks in the event of collision.

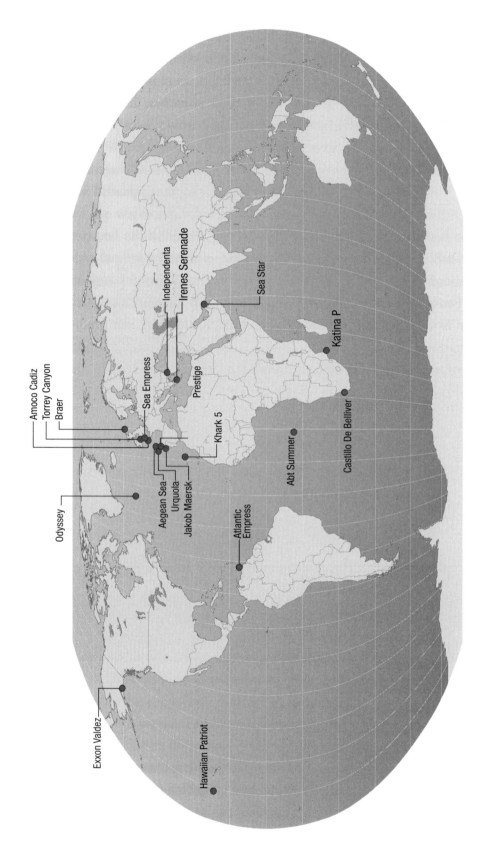

Figure 5.13. *Map showing the location of some of the most notorious tanker disasters.*

In comparison, the *Exxon Valdez*, which ran aground on Bligh Reef in Prince William Sound, Alaska in March 1989 spilled only an estimated 10.9 million gallons (~260,000 barrels) of crude oil. However, the low ambient temperatures, the viscous nature of Alaskan crude oil and the rocky nature of the coastline, with numerous inlets, resulted in extensive pollution in a pristine, wilderness area.

Fortunately, major spills like these are very rare considering the volumes of oil that are shipped around the world each day. Interestingly, the number of tanker incidents has declined significantly in recent years, despite a substantial increase in world oil consumption. Many reasons are cited for this including improved navigation aids, satellite weather surveillance and better communication between vessels, which helps to avoid misunderstandings.

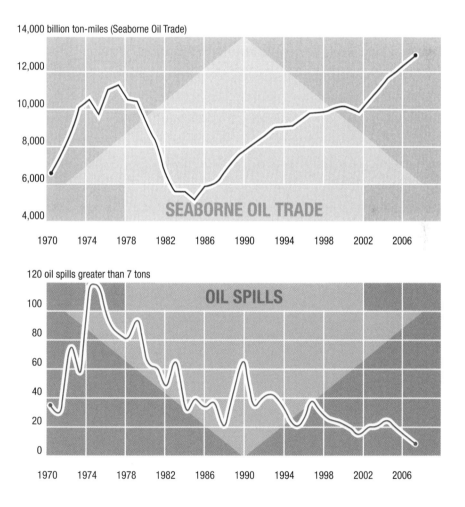

Figure 5.14. Total seaborne oil trade and number of tanker spills over 7 tons, 1970-2008

Today, tankers carry some 2.5 billion tons (~18 billion barrels) of crude oil and refined oil products per year. They are a remarkably efficient and cost effective method of shipping crude adding a mere two cents ($0.02) per gallon to the price of gasoline sold at the pump.[9]

LNG

As noted earlier in this chapter, the vast majority of natural gas is shipped by pipeline but, with a little ingenuity and a lot of engineering, the same gas can be shipped as a liquid, on board special tankers, as so-called LNG (Liquefied Natural Gas). By refrigerating the gas to minus 160°C (about minus 256°F) and pressurizing it very slightly, methane, the principal constituent of natural gas, is transformed into a liquid. It can then be loaded onto an LNG tanker and transported anywhere in the world.

The largest LNG tankers currently have capacities of about 250,000 cubic meters (about 1.6 million barrels) of LNG, equivalent to about 150 million cubic meters (about 5 billion cubic feet) of gaseous product. So, while this is a fairly efficient way to move large volumes of what might otherwise be "stranded gas" (gas that may be uneconomic to produce or too remote to justify a pipeline), it is still much more costly than shipping crude oil. It costs money to refrigerate the gas in the first place and to keep the LNG cold during shipment. The LNG tanks themselves must be made of special alloys to resist the cold temperatures and must be well insulated to minimize losses. Furthermore, LNG has a lower energy density than crude oil so an LNG tanker contains only two thirds the energy of an oil tanker of similar capacity. Finally, on reaching its destination, the LNG must be warmed again, in a process called re-gasification so that it can be delivered to the conventional gas pipeline network for distribution. Thus, LNG is not cheap and currently costs about 5 times more to ship, on an energy equivalency basis, than crude oil.

Also, significant capital investments are required to establish markets for LNG – the expensive refrigeration facilities near the production site, the highly specialized LNG tankers, the re-gasification facilities to convert the product to gas again when it reaches its destination. Despite this, the maritime transport of gas is very much in a growth phase and now represents about a quarter of the total international gas trade. Major LNG exporters today are Qatar, Indonesia, Malaysia, Brunei, Algeria and Nigeria and major importing countries are Japan, Korea, USA, UK and Spain. Not surprisingly, China, too has recently begun imports of LNG.

Figure 5.15. An LNG tanker carries methane gas, liquefied by refrigeration to -160°C, from producing countries to distant markets around the world. The LNG trade has increased dramatically in recent years and now involves some 300 vessels.

CNG

While it is clearly feasible to transport gas across the ocean in a refrigerated liquid state as LNG (and this is the preferred method when shipping distances exceed 2,000 kilometers), it is also possible to ship natural gas in the gaseous state, just as it is, but under high pressure.[10] Compressed Natural Gas (CNG) is exactly what its name suggests - natural gas simply stored under pressure for maritime shipment. This approach offers the benefit of much lower capital costs and significantly lower operating costs than LNG. The gas is compressed typically to pressures of 2,500 to 3,500 psi and stored in steel cylinders (or coiled steel pipe), manifolded together on a customized vessel. The system may, if desired, also feature some refrigeration capability to reduce the gas temperature and, thereby, improve the storage efficiency but, unlike LNG, the gas remains a gas, at all times. CNG carriers have been described as no more than "floating pipelines" and, effectively, that is what they are – long sections of large diameter pipeline, cut into manageable lengths and installed on a ship. The largest such vessel would offer a capacity of around 22 million cubic meters or 800 million standard cubic feet (MMscf). Clearly, this cannot compete with LNG tanker capacities almost

10 times larger but it does offer a reasonable, cost-effective method of transport for shorter distances and smaller loads. Despite several promising commercial possibilities in specific markets, CNG has yet to be applied on any more than a proof-of-concept basis, to date.

Natural Gas Hydrates

Natural Gas Hydrates (NGH) represent another potentially suitable method for the short/medium distance transportation of natural gas. This technology, which has emerged in recent years, primarily from Japan, relies on the natural property of gases like methane to combine with water to form ice-like solids called hydrates (or clathrates) at low temperatures and/or elevated pressures. The process is fully reversible when the hydrate warms up, releasing methane and water. Such hydrates have been known for many years and large quantities are now known to exist in the seabed sediments below 300m depth in many continental shelf locations, where the pressure and temperature favor their formation. Indeed, it has been estimated that naturally occurring gas hydrate deposits represent the largest hydrocarbon gas reserves on the planet, far exceeding those found in conventional gas reservoirs. This, in turn, has raised concerns about the risk of such deposits becoming unstable in a warming world (See Chapter 11) and releasing methane, a powerful "greenhouse gas," to the atmosphere in large quantities.

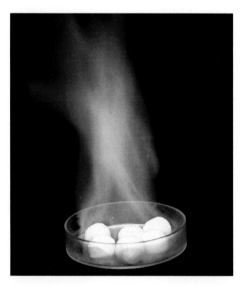

Figure 5.16. Natural Gas Hydrate pellets consisting of methane and water combinedtogether in a relatively stable "ice" burn cleanly when ignited.
(Source: Mitsui Engineering and Shipbuilding)

The use of NGH for transportation purposes, however, does not involve the exploitation of seabed hydrate deposits. Rather, it uses conventional natural gas, from any source, which is combined with water in a cooled, stirred reactor to produce gas hydrate solids. The solids are then separated, compressed into pellets and can be stored and shipped as a conventional bulk solid with very low losses, at temperatures of only -20°C (-4°F). The combined product yields around 180 standard cubic feet (scf) of methane for each 1 scf of hydrate, suggesting the technology is comparable in gross efficiency terms to CNG but certainly cannot compete with LNG or pipeline transportation. However, studies indicate that for regional uses, NGH technology can offer significant reductions, in both capital investment and operating cost, versus LNG and is more flexible and adaptable. Commercial scale-up of NGH is expected in the next few years.[11]

Other Monetization Strategies

While most of the foregoing discussion in this chapter has focused on the physical transportation of hydrocarbons from producer to consumer, there are other methods to monetize difficult-to-access or marginally-commercial hydrocarbon deposits. In the case of gas, in particular, the economics of pipeline or LNG infrastructure can be prohibitive. In such cases, the best solution may be to use the gas at the production site to generate electricity using high efficiency gas turbines. The electricity can provide power for field infrastructure locally and any can be exported and sold to the power grid. Electricity infrastructure is normally cheaper to implement and faster to construct than laying a pipeline and can represent an elegant solution to the problem of stranded resources.

Offshore is a slightly greater challenge but, rather than flaring large amounts of gas as happens today, it could still be used to fire gas turbines and produce electricity. The latter could then be exported to land to supplement often chronically undersupplied power networks via a High Voltage DC trunk line (HVDC). The uses and application of HVDC, and its potential suitability for this type of scenario, are mentioned in Chapter 8. In many developing world countries, access to electricity is poor and, even where available, may be frequently interrupted by power cuts and blackouts. The use of appropriate technology like that listed here could dramatically improve this situation by providing additional power generating capacity while simultaneously reducing waste. Based on satellite observations, the World Bank estimates that a total of 140 billion cubic meters (~4.9 trillion cubic feet) of gas is wasted annually by flaring.[12]

Table 5.2. World Bank estimates of the volume of natural gas wasted by flaring in the past few years, by country

Estimated Flared Volume from Satellite Data			
Volumes in bcm*	2005	2006	2007	2008
Russia	55.2	48.8	50.0	40.2
Nigeria	21.3	19.3	16.8	14.9
Iran	11.3	12.1	10.6	10.3
Iraq	7.1	7.4	7.0	7.0
Algeria	5.2	6.2	5.2	5.5
Kazakhstan	5.8	6.0	5.3	5.2
Libya	4.4	4.3	3.7	3.7
Saudi Arabia	3.0	3.3	3.4	3.5
Angola	4.6	4.0	3.5	3.1
Qatar	2.7	2.8	2.9	3.0
Uzbekistan	2.5	2.8	2.0	2.7
Mexico	0.9	1.2	1.7	2.6
Venezuela	2.1	2.0	2.1	2.6
Indonesia	2.7	3.0	2.4	2.3
USA	2.0	1.9	1.9	2.3
China	2.8	2.8	2.5	2.3
Oman	2.5	2.2	1.9	1.9
Malaysia	1.7	1.8	1.7	1.9
Canada	1.2	1.6	1.8	1.8
Kuwait	2.5	2.5	2.1	1.8
Total top 20	**142**	**136**	**129**	**119**
Rest of the world	**20**	**21**	**19**	**22**
Global flaring level	**162**	**157**	**148**	**140**

*bcm = billion cubic meters

CHAPTER 6
Oil Refining and Gas Processing

Chapter 6
Oil Refining and Gas Processing

Oil Refining

About two-thirds of crude oil produced is used to create gasoline. The remaining one-third is used to create other fuel and hydrocarbon products and, ultimately, a multitude of petroleum-based products and byproducts. Major hydrocarbon products created directly from the refining of crude oil include:

Petroleum gas *(methane, ethane, propane, butane)*
- Applications: Heating and cooking; plastics manufacture; liquid petroleum gas *(LPG)*

Naphtha
- Applications: Solvents and intermediates used to produce gasoline

Gasoline
- Applications: Motor fuel

Diesel and kerosene
- Applications: Motor and jet fuels

Lubricating oils
- Applications: Machine oil, motor oil, greases

Fuel oils *(industrial fuels)*

Solids
- Waxes, tars, asphalt, petroleum coke

Oil refining also provides the feedstocks for petrochemical plants – from which agricultural chemicals and thousands of petroleum products are created – including plastics, ink, paints, polishes, nylons, cosmetics, roof shingles, candles,

Vaseline, tires, auto care products, bug killer, vinyl siding, rugs, garden hoses, crayons, heart valves, and on and on. It is almost mind-boggling to consider how the oil refining industry has evolved – and the intricate and inextricable role it plays in everyday life worldwide.

History gives differing credit for the first refineries – depending on the hydrocarbon origin (coal, tar, "rock oil"). The first refineries were built in the 1850s. Perhaps the first was that commissioned in 1850-51 by James "Paraffin" Young, in Bathgate Scotland. Young is credited with the widespread introduction of kerosene, derived originally from a seep in England but later from coal and oil shale.

Figure 6.1. Dr. James "Paraffin" Young (1811-1883)

A Canadian, Dr. Abraham Gesner, may also be credited with the birth of the refinery industry, at least in America. Gesner developed a method for extracting kerosene (the name he gave to the refined illuminating fuel) from coal and tar. This led to the first refineries in the United States in New York, in the 1850s.

Figure 6.2. Dr. Abraham Gesner (1797 – 1864)

Also, in the mid-1850s, very small refineries were built in what is now present-day Poland. These resulted from the work of Ignacy Lukasiewicz, who developed a method for refining oil from rock (seeps) to kerosene.

Figure 6.3. Ignacy Lukasiewicz (1822-1882)

The first oil refinery used atmospheric distillation to produce kerosene. Naphtha (gasoline) was an essentially useless byproduct of the distillation process. The refinery industry struggled through the second half of the nineteenth century, but the eventual growth of the auto industry in the United States during the early twentieth century fortified demand for oil, and thus refining. By 1911, gasoline replaced kerosene as the primary refined product. A second blow was dealt to kerosene with the invention of the tungsten filament for light bulbs by William David Coolidge (1910).

In 1913, the breakthrough refinery process of thermal cracking was developed. This process enabled greater production of gasoline and diesel fuel per barrel of oil than distillation only. With war, the rapid growth in aviation technology and commercialization, refining became systematized together with oil production and shipping of refined products. The post-war growth of the auto industry fed the progress and growth of refining. New refineries were built and existing ones were expanded. The next breakthrough was in the 1930s with the development of high octane gasoline – which provided superior performance of Allied planes in World War II.

Figure 6.4. William David Coolidge (1873 – 1975)

Figure 6.5. Modern Oil Refinery

The refining business boomed through the 1950s and 1960s. It was not until the 1970s that the industry (and the oil industry, in general) began facing up to environmental consequences, along with the demand for higher fuel economy vehicles. With that, the refining industry has been diligent and successful in developing substantially cleaner burning, more efficient fuels. However, there is a limit to what can be attained with fuel refined from crude oil, especially in terms of clean burning – and the environment – both in perception and reality.

Crude oil, as produced from a well, does not have very much use other than the rare exceptions of very high quality, clean and light crudes which can be used directly in certain fuel applications. Generally speaking, to be useful, crude oil must be separated into lighter and heavier component parts, each purified, and each "fraction" then subjected to various reaction processes – or refinement – leading to the finished commercial products.

Refineries are large scale industrial plants that, for the most part, operate around the clock. Larger refineries typically process hundreds of thousands of barrels of crude oil per day.

There are three phases to the refining scheme:

- Fractional distillation
- Conversion
- Treatment

Fractional distillation takes advantage of the fact that hydrocarbon components in crude oil are of different molecular sizes – and, thus, have different boiling points. Crude oil can be distilled – separating fractions according to boiling point ranges. The temperature in distillation units (towers) is very high at the bottom graded downward to much cooler temperature at the top. The oil thus separates by weight and temperature into fractions.

Conversion is a series of chemical processes taking certain fractions from the distillation stage and converting them into more valuable components. There are three conversion processes:

- Cracking *(breaking larger hydrocarbon molecules into smaller ones)*
- Reforming *(combining components to create different or larger ones)*
- Alteration *(rearranging and modifying chemical structures)*

Treatment includes blending and purifying refinery streams, adding performance additives (such as to fuels), dyes, etc.

A general schematic of the oil refinery processes is shown below:

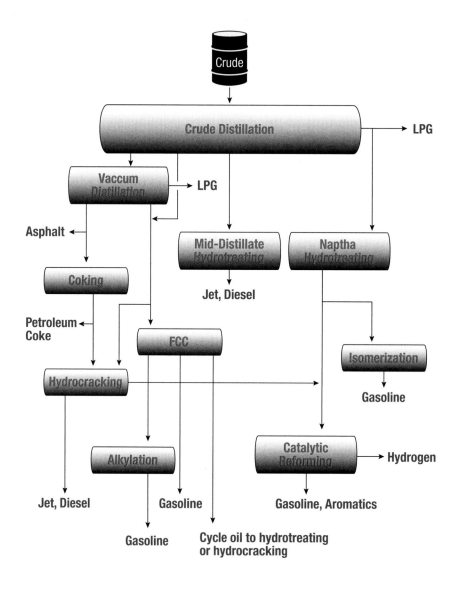

Figure 6.6. Generic Refinery Schematic

Refinery process units and their basic purposes are provided in Appendix D. Together they form the grand scheme that takes one barrel of crude oil and divides it up into its many valuable, hidden pieces.

Refinery Input and Output

Figure 6.7 shows average US refinery "feedstock" (refinery input). Most is crude oil, with the remainder a mixture of refined products (to be further refined or blended with other refinery unit outputs).

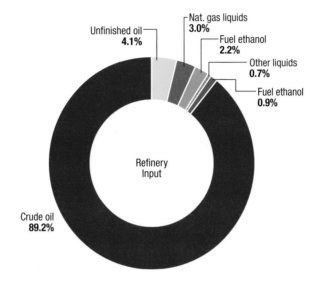

Figure 6.7. Average composition of refinery "feedstock" (Source: EIA)

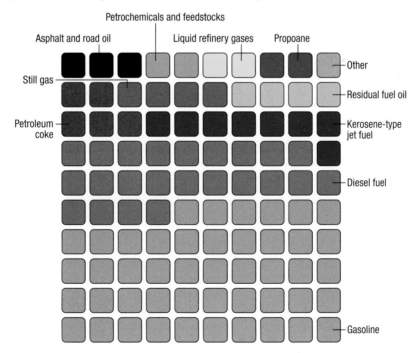

Figure 6.8. Average US Refinery Product Output – From One Barrel of Crude Oil (Source: EIA)

If we take one barrel of crude oil that enters the refinery – we get the collection of refined products listed in Table 6.1. That one barrel (42 gallons) of crude oil once refined typically yields close to 45 gallons of refined products (based on average of all US refinery output):

Table 6.1. What Does One Barrel of Oil Yield in Refinery Products? (Source: EIA)

Products Produced from One Barrel of Oil Input to US Refineries, 2007	
Product	Gallons
Finished Motor Gasoline	19.1
Distillate Fuel Oil	11.0
Kero-type Jet Fuel	3.8
Petroleum Coke	2.2
Still Gas	1.9
Residual Fuel Oil	1.8
Liquefied Refinery/Petroleum Gas	1.7
Asphalt and Road Oil	1.2
Other Oils for Feedstocks	0.6
Naphtha for Feedstocks	0.6
Lubricants	0.5
Miscellaneous Products	0.2
Special Naphthas	0.1
Kerosene	0.1
Finished Aviation Gasoline	0.04
Waxes	0.04
Processing Gain (6.3%)	~2.9
TOTAL	**44.9**

The total volume of products produced from one barrel of oil entering the refinery is greater because of the lower density of the products compared to the oil itself. This increase is called the "processing gain" – and is typically about 6% of the initial volume of crude oil.

A heavy, high sulfur-containing crude oil will be (or can be) refined to a similar end result as a light crude oil with no sulfur – but it takes more processing. A major difference between the two is the output from the first step of distillation. Lighter crudes can have close to 70% light distillates, whereas heavy crudes may contain only half that amount. With that, and the greater difficulty in refining, heavy crudes do not command the same price levels as light crudes. Also, producible crude oil reserves are expected to be increasingly heavy (and sour). Therefore, new supplies of crude oil to refineries will shift slowly but surely to the "more difficult" varieties. The price differences between light, sweet crudes (which will become ever more precious) and heavy, sour crudes will continue to grow.

Figure 6.8 shows the refined product "output" in graphical form. Note that distillate fuel oil is a combination of heating oil and diesel fuel. Thus, gasoline, diesel, and heating oil combine for about 70% of refined product output, on average.

Challenges Facing the Oil Refining Industry

The refinery reaction processes mostly require conditions of extreme heat and pressure, with volatile and toxic reaction byproducts requiring their own processing and control. In times of high refined product (e.g., gasoline) demand – and thus increased profit margins – refineries will operate at high and even maximum capacity. This puts a strain on both the refinery elements and the operating personnel – raising the risk of accident and disaster.

Figure 6.9 shows percent utilization for U.S refineries since 1985. Note the frequency of very high utilization – often exceeding 90%.

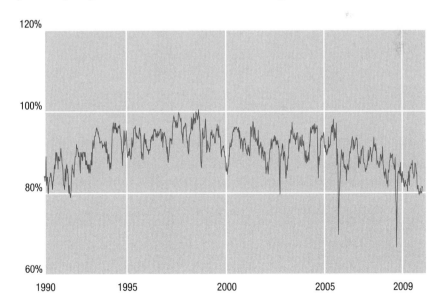

Figure 6.9. Weekly US Percent Utilization of Refinery Operable Capacity (Source: EIA)

The industry has an unfortunate reputation for disaster – with respect to both economic loss and loss of human life. With the general public, this perception often overrides the marvel of engineering and technological advancements that actually characterize refining processes – and the deeply fundamental importance of oil refining to life and progress in the modern world.

"What BP experienced was a perfect storm where aging infrastructure, overzealous cost-cutting, inadequate design and risk blindness all converged. No company should consider itself immune from this kind of disaster." (Carolyn W. Merritt, chairwoman of the BP Texas City refinery accident investigation board, 2005).

Figure 6.10. Wreckage at the BP Texas City Refinery

This statement captures the plight of the refinery industry, especially in the United States. Until the newly sanctioned refinery in Pierre, South Dakota is built, there has not been a new refinery built in the US since 1976. Substantial upgrades to existing refineries have been implemented, but this is not enough to alleviate the long-term strain on capacity.

Refinery environmental and safety regulations have been stringent since the 1970s, when the United States Environmental Protection Agency (EPA) was created. The standards are so tight and compliance is so regulated, that building a new refinery is essentially impossible, without even considering the obstacle of cost. However, the distinct odor of a refinery as is noticed while passing in a car, or of a Texas refinery while passing in a pickup truck, suggests there is still room for improvement with respect to air quality and pollution. Refineries pre-dating the EPA did not take measures to curb air pollution.

In addition to the perception that refineries are ticking time bombs, the continuing question of air pollution, and the extremely high cost it takes to sanction and build a new refinery (a 200,000 barrel-per-day refinery in the US would cost about 4 to 5 billion dollars)[13], the industry is (and has been) shifting more out of the US In fact, the industry has always been, for all intents and purposes, international, as it follows local end product demand.

Table 6.2 lists the twenty-five largest refining companies in the world (as of 2008). The international diversity is apparent.

Also, of the world's twenty largest refineries, only five are located in the United States (Table 6.3). In the modern refining era, the largest refinery for some time was the Abadan refinery in Iran. However, it was substantially damaged during the Iran-Iraq war. While it is difficult to know with certainty which is the largest refinery now, according to Oil & Gas Journal statistics, it appears that it is the Parangua Refining Complex in Venezuela.

Significant additional worldwide refinery capacity is not forthcoming, either. Current global gasoline demand is just below global capacity. The BP Statistical Review of World Energy indicated that world refining capacity for 2009 was nearly 91 million barrels per day – slightly lower than what is required. However, this does not include synthetic fuel (synfuel) refinery output. New refineries anticipated in China and India will provide new capacity, but will not add substantially to global capacity since output will be consumed in the rapidly expanding domestic markets of both countries. Fluctuating refined product demand outside of Russia, Brazil, India, and China, where demand is expected to increase, may relax demand on capacity, at least for periods of time.

Perhaps the increasing US dependence on gasoline imports – and upward pressure on pricing – should be the greater focus of public attention (and not so much on crude oil, exclusively). A smaller number of US refineries are required to do ever more in terms of supplying domestic demand. This has been possible

Table 6.2. The 25 Largest Refining Companies in the World
(Source: Purvin & Gertz Energy Consultants)

Company	Crude Oil Capacity (Barrels per Day)
1. ExxonMobil	5,357,850
2. Sinopec (China)	4,210,917
3. Royal Dutch Shell	3,985,129
4. BP	3,231,887
5. ConocoPhillips	2,799,200
6. PDVSA (Venezuela)	2,642,600
7. PetroChina	2,607,407
8. Valero Energy	2,422,590
9. Saudi Aramco	2,005,000
10. Total (France)	1,934,733
11. Pemex (Mexico)	1,706,000
12. National Iranian Oil Co.	1,660,500
13. Chevron	1,221,900
14. Rosneft (Russia)	1,180,460
15. SK Energy (Korea)	1,106,950
16. Nippon Oil (Japan) (1)	1,099,150
17. Pertamina (Indonesia)	1,035,190
18. Indian Oil Corporation (India)	1,014,000
19. Marathon	974,000
20. Lukoil (Russia)	919,921
21. Kuwait National Petroleum	897,700
22. Flint Hills Resources (2)	815,300
23.Sunoco	775,000
24. Chinese Petroleum Corporation	770,000
25. Motiva Enterprises (3)	762,000

(1) Pre-merger with Nippon Mining; includes affiliates and subsidiaries
(2) Based in Wichita, Kansas; wholly owned by Koch Industries
(3) Joint venture between Shell and Saudi Aramco

Table 6.3. Twenty Largest Refineries in the World (Source: Oil and Gas Journal)

Name of Refinery	Location	Barrels per Day
Parangua Refining Complex (CRP)	Venezuela	940,000
SK Energy Co., Ltd.	South Korea	840,000
Reliance Industries I[1]	Jamnagar, India	661,000
GS Caltex	South Korea	650,000
ExxonMobil	Singapore	605,000
Reliance Industries II[1]	Jamnagar, India	580,000
ExxonMobil	Baytown, TX, USA	557,000
Ras Tanura	Aramco, Eastern Province, KSA	525,000
S-Oil	South Korea	520,000
ExxonMobil	Baton Rouge, LA, USA	503,000
Hovensa LLC	Virgin Islands	495,000
Mina Al-Ahmadi Refinery, KNPC	Kuwait	470,000
BP Texas City	Texas City, TX, USA	460,000
Shell Eastern	Singapore	458,000
Abadan Refinery	Iran	450,000
Citgo Lake Charles	Lake Charles, LA, USA	425,000
Shell Pernis Refinery	Netherlands	416,000
BP Whiting Refinery	Whiting, IN, USA	410,000
BP Rotterdam Refinery	Rotterdam, Netherlands	400,000
Saudi Aramco Yanbu Refinery	Yanbu, KSA	400,000

through expansion and technology advancement. However, refinery closures in the US have reduced the number of operating refineries from 325 in the early 1980s to less than 150 in 2010.[14] This has been a result of both obsolescence and merger activity in the refining industry sector. Placing further strain on the industry is the need to meet low sulfur requirement of gasolines, presently difficult to meet without costly modification, especially to refineries outside of the US.

Figure 6.11 indicates the increase in US gasoline supply – to meet increasing demand. Figure 6.12 shows that US refining capacity has remained relatively flat since 1980, while world refining capacity shows an upward trend.

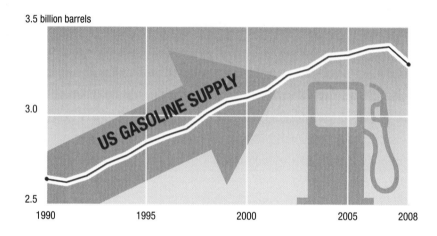

Figure 6.11. US Gasoline Demand – Product Supplied (Source: EIA)

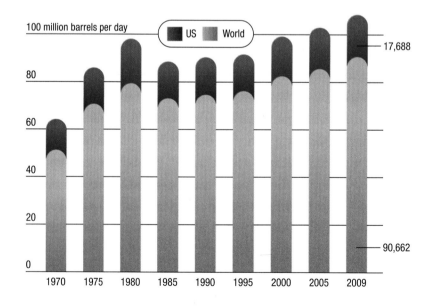

Figure 6.12. US & World Refining Capacity (1970 – 2009) (Source: EIA)

Figure 6.13 shows refinery capacity additions (anticipated) by global region – and the larger relative contributions anticipated in Asia and the Middle East, in particular.

The challenges facing the oil refining industry are quite significant, by any measure. The industry faces growth in global energy and transportation fuel demand, especially in North America and emerging markets in the Asia-Pacific Region. At the same time, new oil discovery is no longer easy, and refinery capacity (and thus supply of refined products) is constrained by a variety of factors, including escalating costs, environmental concerns, and regulatory uncertainty. Return on investment falls into question too, as refinery margins are typically a fraction of crude oil sales margins. Figure 6.14 shows example benchmark refinery margins from three different refining centers. Crude oil margins are always significantly higher.

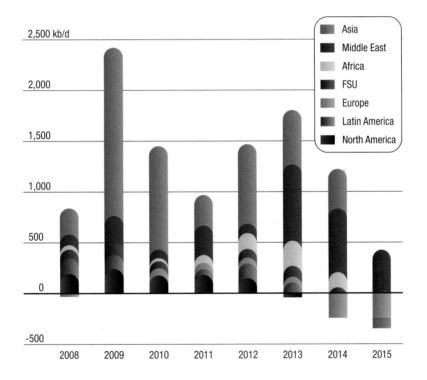

Figure 6.13. Refinery Capacity Additions By Region (Source: The World Refining Outlook, Energy Market Consultants (UK), Ltd.)

Another constraint is the limited availability of a qualified global refinery workforce. And, refinery security cannot be taken for granted. However, the refining industry has always risen to its challenges. It can do so again by building reliable refineries, where possible, and increasing the capabilities and flexibility of existing refineries in handling a greater variety of crude oils and having more stringent impurity and pollutant specifications.

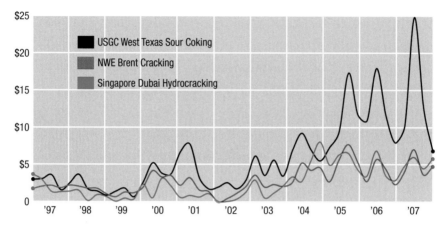

Figure 6.14. Regional Refining Margins, 1996 to 2008
(Source: BP Statistical Review 2010)

Currently, most refineries cannot effectively handle the increasingly sour and lower gravity (heavier) crude oils that are becoming more prevalent. It would take considerable investment to upgrade refineries on sufficient scale to manage such crudes. Among those that can handle the heaviest oil, including that which is recoverable from the Canadian tar sands, and the sourest crudes, are the CITGO refineries in the US The three CITGO refineries that process heavy crudes have a combined throughput of over 750,000 barrels of oil per day.[15] The origin of CITGO is the former American oil company, Cities Service. In 1965, Cities Service re-branded their refining and marketing division, and coined the name CITGO. Cities Service was eventually acquired by Occidental Petroleum in 1982, and through subsequent sales and acquisitions, by 1990 CITGO was a wholly-owned subsidiary of Petróleos de Venezuela, S.A. (PdVSA), the national oil company of Venezuela. CITGO plus other refineries in the US (including those of US companies in which PdVSA holds some ownership share), refine over 30% of all Venezuelan crude oil (from Figure 6.15).

However, the world heavy oil refining picture may be expected to change dramatically in several years time – not to mention the geopolitical ramifications – beginning with an anticipated joint PdVSA – CNPC (China National Petroleum Company) refinery project in the Guangdong province of southern China.

The refinery would be primarily for refining heavy crude oil produced from Venezuela's Orinoco Oil Belt (Chapter 2), starting with a target of 200,000 barrels per day upon completion, anticipated in 2013. The joint objective is to increase China's import of Venezuelan crude oil from current level of over 400,000 barrels per day to one million barrels per day.[16]

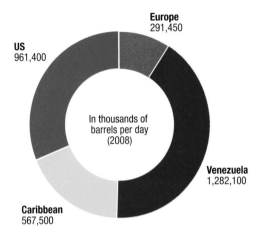

Figure 6.15. PdVSA Oil Refining Capacity, By Region (Source: Oil & Gas Journal)

In addition, PdVSA and Russian oil and gas firms, including state-run Rosneft, Surgutneftegaz, Lukoil, and TNK-BP, are pursuing a consortium that would be in the business of exploration, production, and refining (in other countries) – primarily sourcing the heavy oil reserves in the Orinoco Belt. Venezuela and Japan are also entering agreement on projects for oil production and supply to Japanese markets.

"We are creating a new world, a balanced world. A new world order, a multipolar world. The unipolar world has collapsed. The power of the US empire has collapsed," *he said. "Every day, the new poles of world power are becoming stronger. Beijing, Tokyo, Tehran ... it's moving toward the East and toward the South." (Venezuelan President Hugo Chavez, April 2009)*

In the meantime, and generally speaking, energy efficiency and conservation must be promoted, while energy portfolios are expanded – with greater balance of conventional, heavy, and extra heavy oil sources, conventional gasoline and diesel fuels, biodiesel, compressed natural gas (CNG), liquid petroleum gas (LPG), biofuels, hybrid electrics, ethanol, etc.

Natural Gas Processing

As we know, natural gas provides a relatively clean-burning, readily transportable fuel, heading toward its place as the world's leading fuel source. As we also know, there are enormous global natural gas reserves and opportunities in gas exploration and production to support the expansion and utilization of this energy resource. As discussed in Chapter 2, natural gas reserves include those offshore – in deep sub-sea environments; onshore – including in gas-bearing shales, "tight" (very low permeability) gas-bearing sands, and even in methane-containing coal beds.

The increasingly important natural gas industry segment of the global energy industry continues to evolve, especially following the full deregulation of the US natural gas industry with the Natural Gas Wellhead Decontrol Act (NGWDA) in 1989. It has become extremely competitive and is served by many independent companies and operators within each of its segments – or along its distribution chain:

Exploration … extraction (production) … processing … transport and storage … distribution … end users

The current natural gas distribution chain (or "value chain") is depicted in Figure 6.16:

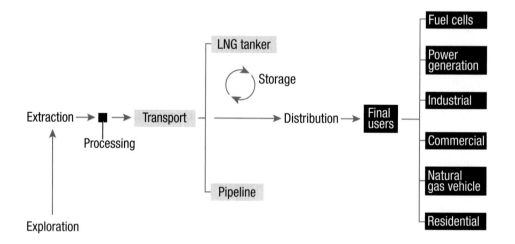

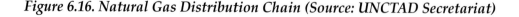

Figure 6.16. Natural Gas Distribution Chain (Source: UNCTAD Secretariat)

Prior to deregulation in the US natural gas industry, this distribution chain was controlled by monopolies. Gas producers supplied natural gas directly to transporters, who in turn provided gas to distributors (wholesalers). Distributors delivered gas to retail end users. After deregulation, the roles were no longer so simple. To appreciate the expansion of roles and of the natural gas industry – note the following statistics from the US Energy Industry Administration (EIA):

- There are over 6,300 natural gas producers in the US alone. These range from large energy companies to one or two person operations.

- There are over 530 natural gas processing plants in the US (there were 727 in 1995). The plants now collectively process 15 trillion cubic feet of natural gas; extracting over 630 million barrels of natural gas liquids (NGLs) per year (1.7 million barrels per day!).

- There are over 160 pipeline companies in the US – operating over 300,000 miles of pipeline.

- There are about 125 natural gas storage operators in the United States – which control around 400 underground storage facilities – with total storage capacity of over 4 trillion cubic feet of natural gas (and average deliverability of 85 billion cubic feet per day!).

- There are nearly 300 companies involved in the marketing of natural gas. Most natural gas supplied and consumed in the US is handled by natural gas marketers.

- There are over 1,200 natural gas distribution companies in the US, with ownership of over 1.2 million miles of distribution pipe.

The remainder of this chapter is devoted to natural gas processing – perhaps the least known and most underappreciated segment of the greater natural gas industry. Natural gas processing is its unsung hero. It is to the natural gas business what refining is to the oil business. However, natural gas processing is simpler in its role. The sole purpose of natural gas processing is to take produced natural gas – separate it into its component parts – and deliver pipeline quality natural gas (essentially pure methane) to the gas transportation system. In the meantime, the separated non-methane components are distributed independently for their own commercial end uses.

It was not until very recently, historically speaking, that natural gas, as produced at the wellhead, could be put to significant use – before which time the commercial natural gas business (and the gas processing industry) of today did not exist. During most of the nineteenth century, "gas" referred to coal gas, produced by retorting coal (i.e. heating it in the absence of air), and used as a source of light – in homes and street lamps (first in Great Britain near the end of the eighteenth century and then in America). Coal gas (also called town gas) was a mixture of hydrogen, carbon monoxide, methane, ethylene and volatile liquids and was, therefore, chemically different from the product we know as natural gas, today. The latter is composed primarily of methane with small amounts of ethane, propane and some other contaminants and, as discussed in Chapter 2, it is generally found either associated with (or dissolved in) oil or in the free state, along with highly volatile low molecular weight, liquid hydrocarbons (condensate).

Originally, natural gas was viewed as a nuisance and a waste product. It could not be effectively transported, or at least not far from its source. Its utility was therefore quite limited, as there was no incentive to develop additional uses. Furthermore, as the nineteenth century was ending, gas lighting was being replaced by electric lights and, consequently gas use, such as it was, went on the decline.

However, there were two developments to point to that triggered the creation of what became the modern natural gas industry. First was that of Robert Bunsen in 1885 – his invention which came to be known as the Bunsen burner. Bunsen discovered that he could mix natural gas and air in such a way as to create a flame that could be used for cooking and heating. As the invention itself and temperature control devices were developed, new use of natural gas in the United States grew rapidly – and followed elsewhere in the world.

With the remarkable increase in natural gas use that ensued, gas pipelines were constructed through the 1920s and into World War II to deliver natural gas, but taking some pause during the Great Depression of the 1930s. It was not until after the war, though, when welding techniques and improved metallurgy and pipe manufacturing methods were developed, that much longer and more reliable natural gas pipelines could be constructed. The post-war pipeline industry in the US boomed for more than two decades. With that, creative uses of natural gas also increased and grew – to include the domestic and industrial uses that are now largely taken for granted. The need for processing natural gas to ever higher quality became necessary – not only for its end uses but to protect the pipelines, as well. For starters, gas had to be free of solids and liquid water to prevent erosion, corrosion or other damage to the pipelines.

Figure 6.17. Robert Bunsen (1811-1899)

Figure 6.18. Natural Gas Processing Plant

Natural gas processing today is not simple, but it is not nearly as complex as oil refining. It also does not have the stringent quality specifications of oil refining, although that is changing, and can be expected to follow the same trend. The relative simplicity is because natural gas does not require so much processing to render it usable.

As mentioned previously, the primary desired end product of natural gas processing is "dry" methane – or "pipeline quality natural gas." Processing, in simple terms, serves to separate the various other gas and liquid "impurities" that are contained in the "raw" natural gas produced from oil, gas, and condensate wells. The impurities in raw natural gas are not waste, but valuable products (or product precursors) in their own right. From Chapter 1, recall that raw natural gas contains predominately methane (CH_4) with nearly all of the remainder including varying amounts of the following:

Heavier hydrocarbon gases and volatiles *(or Natural Gas Liquids – NGL)*:
- Ethane (C_2H_6)
- Propane (C_3H_8)
- Normal Butane (n-C_4H_{10}); Iso-Butane (i-C_4H_{10})
- Pentanes (C_5H_12) and higher

Acid gases
- Carbon dioxide (CO_2)
- Hydrogen sulfide (H_2S)

Other gases
- Nitrogen (N_2)
- Helium (He)

Water *(liquid and vapor)*

Liquid hydrocarbons
- Natural gas condensate
- Crude oil

Mercury (Hg)

Figure 6.19 shows a simplified depiction of the input raw gas and output products of natural gas processing:

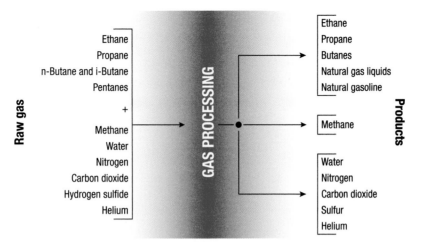

Figure 6.19. Raw Natural Gas to Product Mix

Removing or separating these impurities from methane actually involves only four basic processes:

- Oil and gas condensate removal
- Water removal
- Separation of Natural Gas Liquids (NGLs)
- Hydrogen sulfide and carbon dioxide removal

Some of the gas processing takes place near the production point (wells), but most is transported (through what are known as gas-gathering systems) to processing plants elsewhere – but relatively near the natural gas production.

Of key importance in gas processing is removal of hydrogen sulfide (end product sulfur) and carbon dioxide. Removal of H_2S is the process of "sweetening" (going from sour gas to sweet gas). H_2S can be harmful, and even deadly, if inhaled. At low concentrations, it has an odor of rotten eggs, but with continued inhalation or at higher concentrations (> ~150 ppm), the odor is no longer noticeable due to paralysis of the olefactory nerve (sense of smell). However, H_2S is, in fact, as toxic as cyanide gas (HCN) and concentrations of as little as 1,000 ppm can result in immediate collapse and death from respiratory arrest.

Sour gas can also be quite corrosive to pipelines. The H_2S removal process is also similar to the dehydration process. In this case, the "remover" is a chemical solution that has an affinity for sulfur. Once processed, the gas is essentially sulfur-free. The chemical solution used to remove H_2S can be regenerated (as can the desiccants utilized in the dehydration and NGL removal processes). Solids such as iron (in sponge form) are also used to remove H_2S (and CO_2). Once the H_2S is removed, elemental sulfur is recovered through the Claus process, which is an extraction of sulfur from the separated H_2S solution. The bright yellow solid material often seen piled at gas plant locations is elemental sulfur.

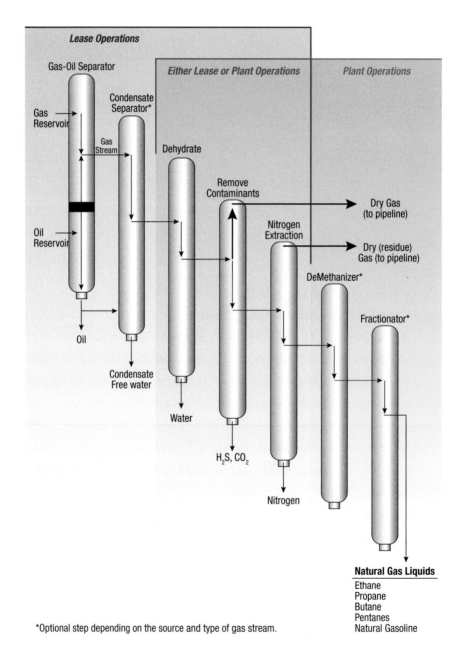

Figure 6.20. Basic gas processing schematic (Source: EIA)

Figure 6.21. Elemental sulfur byproduct of natural gas processing

Additionally, if mercury (Hg) is present in gas, it is removed with molecular sieves (zeolites) or activated carbon in a mercury removal unit.

Besides the removal processes, and prior to them, heaters and scrubbers are installed at or near the wellhead. Scrubbers are to remove solids (such as sand produced with fluids from the wells) and heaters are employed to maintain produced gas temperature that will prevent gas hydrates from forming. Gas hydrates are ice-like crystalline solids within which natural gas is trapped. Hydrates can be most troublesome in restricting gas flow wherever they form in the gas production to transport system in valves and pipe. As mentioned in Chapter 2, naturally-occurring gas hydrates (clathrates) represent a potentially huge, future natural gas resource. However, in process flow lines, their formation can be induced by sufficiently low temperature and high pressure (through gas expansion and associated cooling). This phenomenon also occurs in deepwater gas wells where temperatures can be low (near freezing) at or near the seafloor, at the wellhead of gas wells as gas expands; as well as in pipelines (in the pipes and valves) – restricting flow. Besides increasing temperature, chemicals that break gas hydrates or inhibit their formation can be utilized. One is methanol; others are "anti-freeze" chemicals such as monoethylene glycol (MEG) and diethylene glycol (DEG). Newer methods include surface active agents (surfactants) that disrupt the ice-like crystallization process, thereby preventing gas hydrate formation.

Separated Natural Gas Component End Uses

As mentioned, gas processing serves the sole purpose of separating methane and the components making up raw natural gas. For more on that, see Appendix E.

Besides its fuel uses, methane is also used in the manufacture of ammonia and methanol. The non-methane components have commercial uses, some of which are as follows:

- Ethane has fuel application (in combination with methane) and is used in the production of ethylene, in turn used to produce polyethylene.

- Propane enjoys a variety of uses – primarily residential and petrochemical. Also, an ethane-propane mix is sometimes separated from NGL. The ethane-propane mix is transportable by pipeline for use as refinery feedstock, for example.

- Butane also has a wide variety of uses. Apart from being a valuable fuel for cooking and heating, it can be used as a feedstock for refinery gasoline production, as well as a propellant in aerosols.

- Natural gas liquids (NGL) include ethane, propane, butanes, and natural gasoline (condensate). These NGLs have a variety of different uses, including for enhanced oil recovery (EOR) injection processes in mature oil fields, fuels, and raw materials for oil refineries and petrochemical plants.

- Natural gasoline is a mixture of light hydrocarbons, starting with pentane. A primary use of natural gasoline is blending into refinery gasoline.

- Elemental sulfur is used in the production of industrial sulfuric acid, rubber vulcanization, and manufacture of black gunpowder.

- Helium has many specialized uses – beyond filling party balloons. These include rare document preservation (The United States Declaration of Independence is preserved under a helium blanket); pressurizing agent for liquid fuel rockets; leak detection; inert gas shield for arc welding; cryogenics; cooling medium for nuclear reactors; mix gas for deep water "saturation" diving, and many others.

Just as with oil refining, natural gas processing is a marvel in many respects from a chemical and engineering standpoint. Like oil refining, gas processing is the key link between a raw material that is of limited use as is – and a host of commercial end products that can be generated from it. And the progress that has been made in natural gas processing in a short time is also amazing – considering that the commercial natural gas industry began less than 100 years ago, and did not have the beginnings of its present infrastructure until the 1950s. The most favorable aspect of natural gas processing relative to oil refining is that the adverse effects of oil refining are largely non-existent in natural gas processing. Gas processing is of lesser complexity than oil refining. Gas processing facilities are also of substantially lower cost than new refineries, more environmentally "benign," and expansion of capacity and capability (including in the US) is thus more realistic, worldwide.

These all favor the transition from crude oil as the predominant energy source to natural gas.

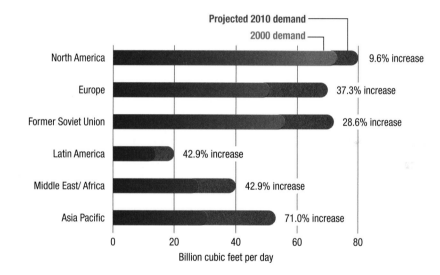

Figure 6.22. Projected Increase in Natural Gas Demand (Source: ExxonMobil)

Further shifts will occur with new coal technologies and alternative energy sources, which are discussed in the next chapters.

CHAPTER 7

The Role of Coal

Chapter 7
The Role of Coal

Figure 7.1. Coal-burning Power Plant

Clean coal's like healthy cigarettes – it does not exist (Al Gore, September 2008)

Such sweeping statements, driven by a doctrinaire view against all fossil fuels, along with images such as that in Figure 7.1, have muddled the public debate on energy. Furthermore they are an affront to the tremendous historical importance of coal's contribution to society – and its key future role in meeting the energy needs of a growing world as the industry strives to accommodate political and social demands. It cannot be forgotten that coal, as a much better alternative to wood, served as fuel to the steam engines that essentially drove the Industrial Revolution – following the development of deep shaft coal mining in the UK in the late 1700s. Coal thus became the primary source of energy for domestic use, industry, and transportation from that point on until the 1950s. Since then, the combination of oil and natural gas has replaced coal as the lead energy source. However, nearly 25% of energy needs in all sectors are still met by coal (Figure 7.2) – and its relative contribution is now on the rise. China, the emerging world economic superpower still draws about 70% of its primary energy supply from coal.

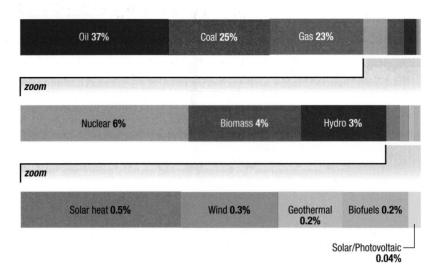

Figure 7.2. Global energy source contributions

Historically, coal has had many uses. For example, the gas in early street gas lights in most cities was from coal. In the nineteenth century it provided fuel to boilers on steam-powered trains. Coke, a byproduct of coal processing, became the primary fuel used in steel manufacture. Around the turn of the twentieth century, coal's use in household heating proliferated. Today, in developed countries, coal finds its primary use in generating electricity. In the US, which holds one-quarter of the world's coal supplies, about 85% of its coal is used for electricity generation. And about half of the electricity used in the US is generated from coal. Globally, coal accounts for about 40% of electric power generation, as coal in developing countries has a greater variety of uses (see Chapter 8).

Apart from electrical power generation, another major use of coal is in the production of coke. There are thousands of byproducts of coal, including those common to households and personal lives, many of which are depicted in the "Coal Tree" below.

The coal industry has had to contend with two broad areas of social and political concerns. First it had to move away from dangerous and unhealthy mining practices. These included the almost always present danger of underground explosions because of the presence of adsorbed natural gas (the latter in the flip, positive side, is what is produced as coalbed methane, CBM, from other coal formations). The second challenge is to effect less polluting, cleaner-burning coal. Coal generates ash which, if allowed to enter the atmosphere as was the case decades ago, obscures the atmosphere in the vicinity of the plants and causes breathing problems and ailments. Other coal combustion byproducts are nitrogen oxides (NO_x) and sulfur oxides (SO_x) which if released to the atmosphere will generate among other problems, "acid rain," a serious environmental concern. In the West, controlling ash and NO_x and SO_x emissions has been central to coal management.

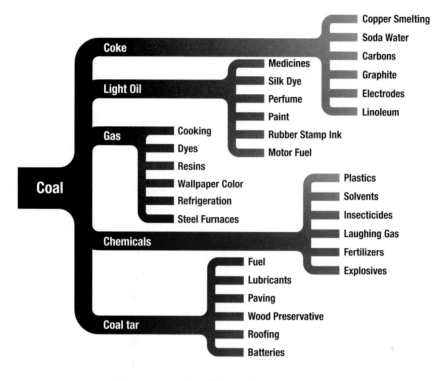

Figure 7.3. Coal Tree (Source: USGS)

China, with very large coal reserves, has depended on coal for decades. Without the necessary attention to ash and other emissions, this has led to severe environmental problems to the point that blue sky is rarely seen in many Chinese cities. Along with effluent water contamination, which has been introduced into surface water bodies, this has resulted in 20 of the world's 30 most polluted cities. Things have improved during the last decade but China still has a long way to go because remedial action is affected by the country's constant demand for additional power generation. At the time of writing, China has been adding on average one and one half commercial coal power plants per week. China has been under pressure from international environmental groups with respect to cleaner coal-burning practices. However, even under the scrutiny and increasingly stringent enforcements, coal remains inexpensive in comparison to other energy sources and is still relatively easy to extract.

It may come as a surprise that coal is the most rapidly growing fuel, based on global usage – and that has been the case for the last several years. Proven, or economically recoverable, coal reserves are trending downward although the amount is significantly greater than proven reserves of oil and gas. Coal comes at relatively low cost, and until recently, with an absence of price volatility. Figure 7.4 shows coal and natural gas prices relative to oil price (West Texas Intermediate crude) since 2000. Even with the recent uptick in price (due to the sharp increase in Asian demand) and factoring in a possible future additional "carbon cost" – coal's cost per unit of energy will still be considerably less than either oil or natural gas – as it will continue to dominate base electric power production in both the developed and developing worlds.

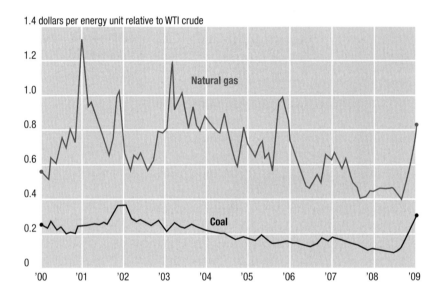

Figure 7.4. Coal and natural gas price
(dollars per energy unit) relative to WTI crude oil (Source: EIA)

Economically recoverable coal reserves exist in more than 70 countries.[17] Six countries – the US, China, Russia, India, Australia, and South Africa – account for over 80% of the global coal reserves (and about the same share of coal production). As can be seen in Figure 7.5, the picture of which holds and controls global coal reserves is quite different in comparison to oil and natural reserves, as we saw in Chapter 2. The Middle East is insignificant with respect to coal reserves, while the US, the former USSR., and China are the most prominent – and self-sufficient.

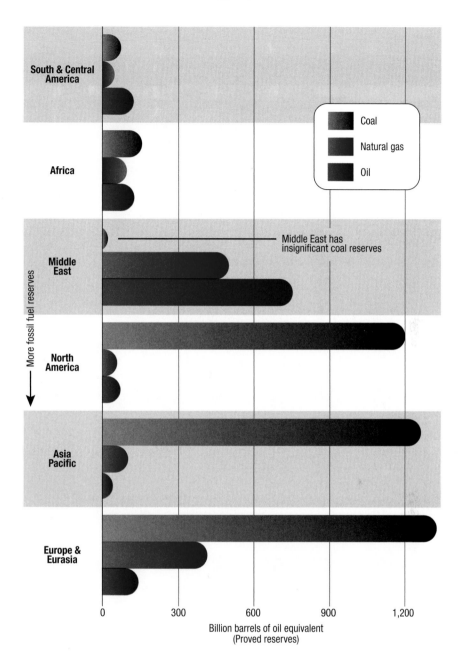

Figure 7.5. Hydrocarbon (and Coal) Reserves in Global Regions
(Source: World Coal Organization)

The largest coal market is Asia, which consumes over 55% of global coal supply (with China accounting for most of that). The rate of increase in the production and consumption of coal in Asia relative to the rest of the world is staggering (Figures 7.6 and 7.7).

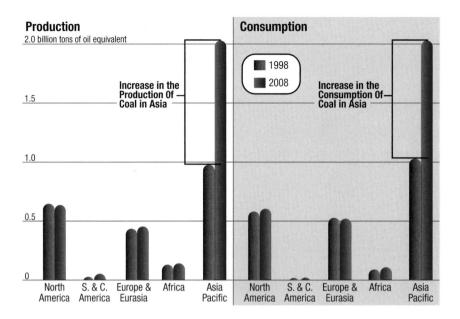

Figure 7.6. Coal Production and Consumption by Global Region (1998 and 2008) (Source: BP World Energy Review 2009)

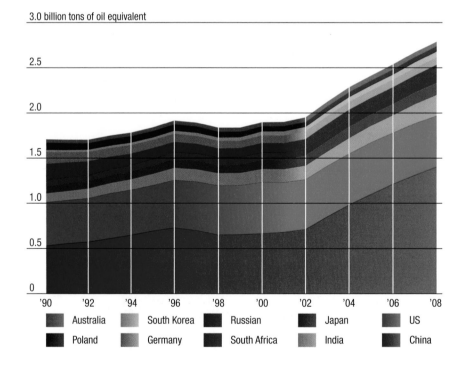

Figure 7.7. Coal Consumption – 10 Largest Consuming Countries (86% of Global Total) (Source: energyinsights.net)

Future of Coal

Figure 7.8 provides one projection (from the EIA) of the relative contributions of different fuel sources to the world market energy usage. Note the steeper rise in the contribution from coal, already in effect.

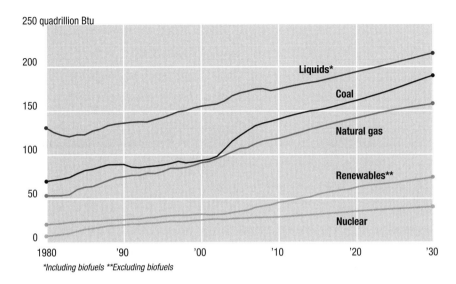

Figure 7.8. World Marketed Energy Use by Fuel Type – 1980-2030 (Source: EIA)

With that in mind, there is also coal's contribution to carbon dioxide (CO_2) emissions, shown by fuel type in Figure 7.9. The role of coal in this regard is the basis for its vilification and the pressures to "clean up its act" in a carbon-constrained world.

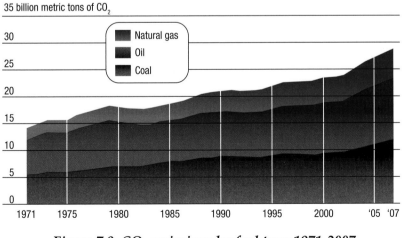

Figure 7.9. CO_2 emissions by fuel type 1971-2007
(Source: IEA)

But in the meantime, the world's economy, fed by the expansion of the massively populated countries such as China and India, is not going to stop growing. There is, and will be, an increasing need for coal, with its otherwise favorable characteristics, to supplement the growing need for energy. It will further stand to reason that developing countries, as they are already doing, will turn to coal as an economically viable source of energy to power their expansion.

While there are misperceptions and irresponsible portrayals of coal by its detractors (and beneficiaries of clean coal and carbon tax legislation) – its "dirtiness" and related shortcomings are well-documented, as are its actual advantages and disadvantages – politically, socially, and industrially. Regardless, the direction in which coal must be taken is set. So, it is very important to develop clean coal technologies, independent of the ultimate reality of desired decarbonization targets.

The notion of "clean coal" is not new. Since before the Industrial Revolution, there has been public outcry over not only coal mining and labor conditions, but with respect to the environment – the waste products, smoke, eventually acid rain and smog, and now climate change. Therefore, for example in the US, coal-based electricity providers have invested over $50 billion in technologies throughout the last 30 years to reduce emissions. Today's under-fire coal industry is substantially "cleaner" than it ever has been. There are now about $12 billion invested in clean coal research in the US, in 43 states.[18] Combined energy company and government programs are focused on development, demonstration, and eventual implementation of technologies that may enable near-zero emissions and capture and store greenhouse gases – a lofty, and to say the least, challenging goal.

There is a variety of so-called clean coal technologies employed in the combustion of coal to generate fuel gases – all with the objectives of producing lower waste products (particulates, emissions of gaseous pollutants and/or greenhouse gases, especially carbon dioxide) – to meet the increasingly stringent environmental standards and anticipated future requirements. In today's terminology, different methods fall under two general categories – Traditional Clean Coal Technology and Truly Clean Coal Technology. Traditional Clean Coal Technology is that which removes coal impurities to allow more carbon and oxygen to react when the coal is burned (increasing efficiency), as well as after burning, through filtration of ash and polluting gases such as sulfur dioxide (SO_2) and nitrogen oxides (NO_x) from the emissions. Traditional technologies are thus pointed to reducing air pollution, and acid rain. In the meantime, high sulfur-containing coal producers have suffered, while low sulfur-containing coal producers have thrived, comparatively. With modern emission removal processes, coal producers may claim "emissions-free" coal burning, and that may be the case in the "traditional" sense. But, that does not include CO_2 and methane (and other carbon-containing gases) – greenhouse gas emissions.

Truly Clean Coal Technologies are those which target significant reduction in greenhouse gas emissions. Among the most prominent clean coal technologies under development are: Underground Coal Gasification (UCG), Carbon Capture and Sequestration (CCS), and Integrated Gasification Combined Cycle (IGCC). These three are briefly described below.

Underground Coal Gasification (UCG)

UCG is a method of converting coal while it is still in the ground into combustible gas which can then be used for industrial heating, power generation, or for the supply of hydrogen, synthetic natural gas (syngas), or diesel fuel. The simplest UCG process uses two wells, set some distance apart, drilled into the coal to desired depths. One well is used to inject an oxidant mixture (water + air or water + oxygen) that is capable of combining to burn coal and the other well is used to produce the gas mixture product. In the UCG process, the coal is not actually allowed to burn, though. The injection of oxidants is controlled so that the temperature rises to a point at which gases (syngas) separate from the coal. The continued flow of the oxidant stream during injection generates and displaces the gas to the producing well. In a more complex process, injection boreholes can be drilled into the coal seams so that the injection point can be changed over time – further increasing recovery of syngas from the coal. Since conventional coal extraction methods leave a majority (nearly 85%) of coal resources inaccessible, UCG provides a method to enhance reserve recovery several-fold, as it can reach depths well beyond the safe and cost-effective mining capabilities.

UCG is not new. Sir William Siemens considered the idea before 1870 – for the elimination of unusable coal waste. The great Russian chemist, Dmitri Mendeleyev, the creator of the Periodic Table of the Elements, developed the underground coal gasification concept further, over the next two decades. Eventually, the 1904 Nobel Prize recipient Sir William Ramsay continued on, overseeing the first experimental work which was conducted in the UK prior to

World War I. However, further effort was delayed until after World War II, when Russia revived research based on Ramsay's work. By the 1960s, there were 14 underground coal gasification plants in Russia.

Figure 7.10. Fathers of Underground Coal Gasification: From Left to Right – Sir William Siemens (1823-1883); Dmitri Mendeleev (1834-1907); Sir William Ramsay (1852-1916)

The discovery and exploitation of huge natural gas resources led Russia away from underground coal gasification. Today, there are still a limited number of active UCG plants worldwide. Only recent advancements in technology have made UCG economically viable on a grander scale, and thus much more attractive. This is especially the case in China, the largest consumer of coal in the world. The Chinese government funds the largest UCG development programs. There are about 30 such projects in different stages of development in China. The main advantages are low plant costs and no coal transport costs – especially desirable in the emerging markets of not only China but, India, Australia, and South Africa – countries in which projects are underway. In fact, Sasol and Eskom of South Africa both have UCG pilot plants in operation. UCG demonstration projects are also underway in the US, Europe, Japan, Indonesia, and Vietnam.

An intriguing feature promoted by UCG proponents is the potential synergy with Carbon Capture and Storage (CCS). Captured CO_2 could be stored in the underground coal (injected in the UCG wells) after the gasification is complete. Critics of UCG cite the greater production of CO_2 per unit energy, and the waste products of coal combustion that are left behind (below in the underground coal) can leach pollutants into nearby groundwater.

Carbon Capture and Storage (CCS)

CCS is a process whereby carbon dioxide emitted from industrial plants (including coal power plants) is captured, injected into geological formations underground, and stored there (see Chapter 11). Nearly $3.4 billion were allocated to CCS in the 2009 US stimulus package, and the European Union has established incentives for power plants to adopt CCS. It is widely believed that in order to meet the targets of reduced emissions of greenhouse gases, CCS will be the required technology.

A very limited version of CCS has already been employed. There are years of operational experience of underground injection of CO_2 for enhanced oil recovery (EOR), and technologies analogous to CCS, such as acid gas injection and natural gas storage. However, the scale of required CCS for the management of CO_2 emissions is orders of magnitude larger than the EOR experience.

There are nearly 100 projects at various stages of development, eight of which are operational (in Algeria, Canada, France, Japan, and Norway). The Sleipner project in Norway has employed CCS since 1996. This project is operated by Statoil-Hydro. The Sleipner field produces natural gas containing up to 9.5% CO_2. The CO_2 is separated and re-injected into a sand layer (Utsira formation) above the gas-producing zone, but still over 3,000 feet (or about 1,000 meters) below sea bottom. Statoil-Hydro thus avoids payment of a carbon tax, which is applied in Norway.

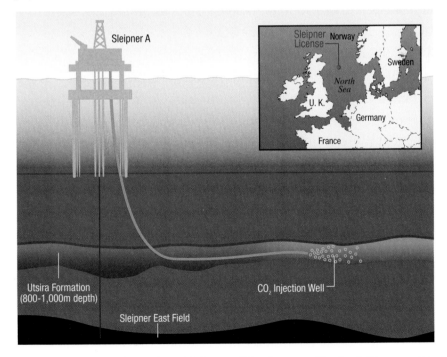

Figure 7.11. Diagram based on Sleipner CO_2 Storage Project (Source: Statoil)

For broad and significant application of CCS – beyond special situations – a key step that must be taken is the integration of large scale CCS systems into commercial-scale power plants. Recent work suggests that CCS at the scale that would be required for these applications does not seem to be possible. Injection into a closed system which is fundamental to CCS would require far larger pore volumes that have been postulated earlier. This does not bode well for CCS at any cost.

Integrated Gasification Combined Cycle (IGCC)

IGCC combines coal gasification and a combined cycle process. The gasification process takes coal and partially combusts it with oxygen and steam to form syngas (carbon monoxide and hydrogen). The syngas is then cleaned to remove particulates, mercury, and other emissions such as sulfur dioxide. In the combined cycle process, a gas turbine is driven by the combustion of syngas produced from the coal gasification process. The exhaust gases are heat-exchanged with water and steam to generate superheated steam – which is used to drive a steam turbine before it goes on to drive the gas turbine. IGCC provides high system efficiency (consuming less fuel per kilowatt-hour of output than conventional generators) and ultra-low pollutant levels. More important to clean coal proponents, the IGCC technology is seen to offer the potential means to capture and store CO_2.

And then there is FutureGen – a US Department of Energy (DOE) project to build a coal-fueled, near-zero emissions power plant. The project was previously pulled in January 2008 during the Bush administration because of increased cost concerns (budgeted at $950 million – but reportedly inflating to $1.8 billion). The project was revived in 2009 under the Obama administration, with over $1 billion earmarked through the stimulus bill – to be combined with about $600 million in private industry funding. No new funding was sought by the DOE for 2010. FutureGen endeavors to combine and test several new technologies in a single plant, including coal gasification and hydrogen production (IGCC), emissions controls, electricity generation, and CO_2 capture and storage (CCS). The diagram below illustrates the FutureGen process – the goal of which is a near-zero emissions power plant.

The FutureGen facility is anticipated to be capable of generating 275 megawatts of electricity, enough to power about 150,000 average US homes. Mattoon, Illinois was selected as the site.

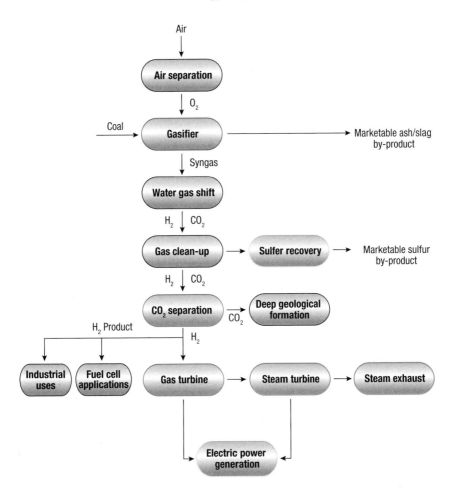

Figure 7.12. FutureGen Process Diagram (Source: FutureGen Alliance)

Figure 7.13. Artist rendering of the US Department of Energy FutureGen site (Source: Wired Science)

These Truly Clean Coal technologies have not yet been developed to the extent that they can be commercially cost-effective, or at least competitive with nuclear power technology, for example. Make no mistake about it, though, coal is here it stay. How clean it can be remains to be seen – as it continues to establish its foothold and prominent role in generating electricity to a growing world.

The next chapter follows into **Power Generation.**

CHAPTER 8

Power Generation

Chapter 8
Power Generation

In assessing the level of development of countries or societies, several metrics may be used. The level of sanitation and access to running water are examples of such metrics. At a somewhat higher level is a country's capacity to generate and distribute electricity to its inhabitants and how efficiently that energy is put to use. Of all end-use energy sources, electricity is arguably the most versatile. It can be generated centrally, stepped-up to high voltage by transformers and transmitted across vast distances with minimal losses, then stepped-down to normal voltages, delivered directly to consumers, and used for an apparently endless list of purposes. Conversely, it can be generated locally, with relatively simple equipment, used as needed, and any excess can be sold back to the central power grid.

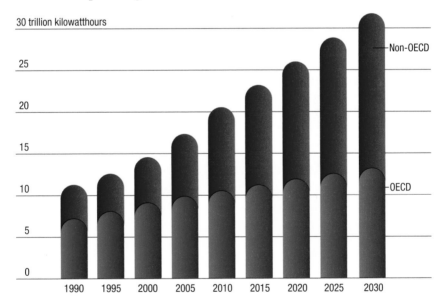

Figure 8.1. World Electric Power Generation, 1990-2030 (Source: IEA, EIA)

As indicated in Fig. 8.1, from 2005 - 2030, global electricity demand will approximately double. The majority of that demand growth will take place in the developing world, primarily in China, India, Indonesia, etc. In order to meet the increased demand, capital investment in excess of $11 trillion will be required. Approximately half of that will be invested in the developing world while the other half will be spent replacing and upgrading ageing generators and transmission/distribution infrastructure in OECD countries. Investment opportunities in the power sector will abound but they will require careful screening due to the widespread use of subsidized tariffs and the non-reliability of sovereign debt in many developing countries. This chapter discusses power generation, in its historical context, as well as with respect to future trends – the implications of which can be investigated further in the interest of wise and forward-thinking investment.

In general, and as mentioned previously, coal is the major hydrocarbon fuel used for electricity generation, accounting for about 40% of world electricity supply but ranging anywhere from less than 20% to over 80%, depending on the country. Natural gas, which contributes 20% of global electricity production, is the second major fuel while hydropower and nuclear both command about 16% each. Oil is used only for a little over 6% of world electricity supply and this is predominantly in countries in the Middle East or those with poor electrical grid infrastructure. With the exception of very remote areas or isolated islands, in the developed world and, even emerging China, there is very little power generation from oil. The remaining balance of less than 2% comes from the combustion of waste or biomass, geothermal, wind and solar.

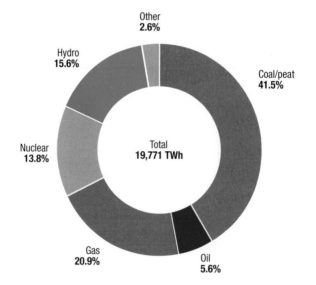

Figure 8.2. World electrical supply (Source: IEA, 2008)

This mix, shown in Figure 8.2, is expected to change significantly over the next 25 years. Coal will remain the primary fuel for electricity generation but both gas and nuclear power are expected to increase their share, depending on gas prices, climate change issues and public acceptance of the need for additional nuclear capacity. Oil use for electrical power generation will fall further while hydroelectric will increase, although its absolute overall share will decline. Some suggest that wind and solar will experience significant growth, providing, on aggregate, perhaps 7% to 8% of world electricity by 2030. Others point out that these sources without massive government subsidies are not viable and that coal, natural gas and nuclear will continue to dominate at roughly the same overall share in the foreseeable future.

In some countries, however, there have been claims and ambitious plans to derive 20% of electrical power generation from wind power alone. These may sound like unattainable projections, considering where these technologies are today (currently far less than 1%). The trouble with some of these projections is that they are often published by vested interests which are prone to be highly misleading. For example, there is a constant (and perhaps deliberate confusion) between installed capacity and actual output into the electrical grid. The trouble with wind is that it is intermittent and good prospects are far away from demand centers. A 25% load is considered quite good. In Texas where at the time of this writing 9,600 MW of wind capacity were installed, on February 2, 2010, for example, only 527 MW of wind was entering the system for only 5.6% of installed capacity.

In Denmark wind accounts for almost 20% of installed capacity. However, Denmark has the opposite problem. While the wind blows it cannot absorb all the power, and since it cannot be stored, the country exports electricity to neighboring Sweden. Other European countries with wind development include Spain, with 10% and Germany, with 7% of installed power capacity.

In March 2010, the International Energy Agency, in cooperation with the Organization for Economic Cooperation and Development's Nuclear Energy Agency, released a study called "Projected Costs of Generating Electricity." The results showed clearly that wind was the least costly source of new generation. However, using what the study called the "levelized costs of electricity," a metric that includes key factors such as the discount rate, construction costs, load factors, fuel prices, and carbon costs, it was found that nuclear power is the least expensive option for new generation in North America, Europe, and Asia Pacific. Meanwhile, wind energy was often the most expensive option regardless of location.

Solar photovoltaic power is much more expensive but it too has fallen from around $20 per kWh to about $2.70 per kWh, and new thin-film techniques promise to reduce these costs by another order of magnitude. Even in such case it would still be several times more expensive than nuclear, coal or natural gas.

Historical

The modern power industry has its roots in the work of nineteenth century scientists like Michael Faraday, who first demonstrated electromagnetic induction and James Clark Maxwell, who studied Faraday's work and first defined the laws of electromagnetism. Building on such work were Thomas Edison who developed among other things the ubiquitous (but, perhaps, soon to be extinct) electric light bulb and Nikola Tesla, who invented and patented the brushless alternating current motor in the latter part of the 1880s.

Figure 8.3. Thomas Edison Figure 8.4. Nikola Tesla
Two of the early pioneers of commercial electricity

George Westinghouse, an American engineer and entrepreneur was a key figure in the industry's development helping to piece together the work of many into the functional system of alternating current (AC) generators and transformers that constitute the components of a modern day electrical grid. The first electrical distribution networks used low voltage direct current (DC) but these were replaced by the more efficient AC systems championed by Tesla and Westinghouse. This was a subject of a prolonged and bitter dispute with Edison who supported the use of DC for general electrification.

The first commercial generation and distribution of electrical power was from Edison's Pearl Street Station in New York, opened in 1882, using DC, as noted above, and supplying only 59 consumers. However, the system was inefficient and limited in its coverage due to power losses in the lines, a natural result of using low voltages. Consumers had to be sited within a mile or so of the generating station.

Working for Westinghouse, who had purchased the rights to his patents, Tesla scored a major success when his design won the Niagara Falls Commission contract in 1893 – for the generating plant and transmission system from the world's first major hydroelectric facility at Niagara Falls. That same year, Westinghouse and Tesla were given the opportunity to demonstrate the use of AC to light up the World's Fair in Chicago. By late 1896, power generation from the Niagara Falls plant was underway. The success of this project was a turning point for the acceptance of AC as an efficient means to generate, transmit and distribute electrical power. It was quickly emulated by others and electricity became one of the fastest implemented technical advancements, in terms of mass access, in mankind's history. By 1900, about 8% of American households had electricity, and by 1930, over two-thirds. Today, the number of American households with access to electric power is greater than 99%.

Figure 8.5. George Westinghouse

Basics

Electricity is essentially the flow of electrons in a conductor. Each electron carries energy and the total energy conveyed is, therefore, related to the total number of electrons that pass along the conductor. This number, in turn, depends on the size (i.e., cross-section) of the conductor, what it is made of, and the potential difference between its ends (the voltage). The larger the voltage, or the thicker the conductor, or the more conductive the material used, the more electrons can pass through in a given period of time and, therefore, the greater is the power transferred. Along the way, a proportion of the energy is dissipated by parasitic effects due to factors like heating (caused by the resistance of the conductor), the generation of magnetic fields, etc.

Creating the flow of electrons in the first place is the process of power generation and there are many methods that can be used to accomplish this. Today, the vast majority of electrical power is generated by thermal means, which involves the use of some other fuel to heat and vaporize water which, in turn, is used to drive steam turbines. This is the process used by those power stations that burn coal, gas, oil or use nuclear sources to generate the heat needed to produce steam. It is also the basis of geothermal electricity generation, which uses natural heat from geological sources to produce steam. Alternative non-thermal methods to produce electricity include hydropower and wind – both of which extract energy from the flow of, respectively, water and air, through turbines, and photovoltaic ("solar") systems, which use silicon or other semiconductor panels to convert a portion of the sun's incoming ultraviolet light directly into electricity, by a phenomenon known as the photoelectric effect.

Balancing Supply and Demand

Electricity demand constantly changes, both throughout the day and throughout the year. The changes are based on typical diurnal patterns of human activity and on seasonal climate change, superimposed on these patterns. Thus, while all regions exhibit higher demand during work hours or when people cook or heat their homes during leisure times, such changes are more pronounced in latitudes where winter's effect is felt or those where the luxury of air-conditioning can alleviate the heat and humidity of the sub-Tropics.

Such gross changes in power demand are fairly predictable and can be planned for adequately by adjusting generating capacity, accordingly. This background average requirement constitutes the so-called base load and is provided by the vast networks of coal, hydro and nuclear power stations of developed countries (or trading blocs). Short term fluctuations in demand are dealt with by using more dynamic power generating options (e.g., the use of gas-fueled facilities). As demand increases, more stations are brought on line to cope.

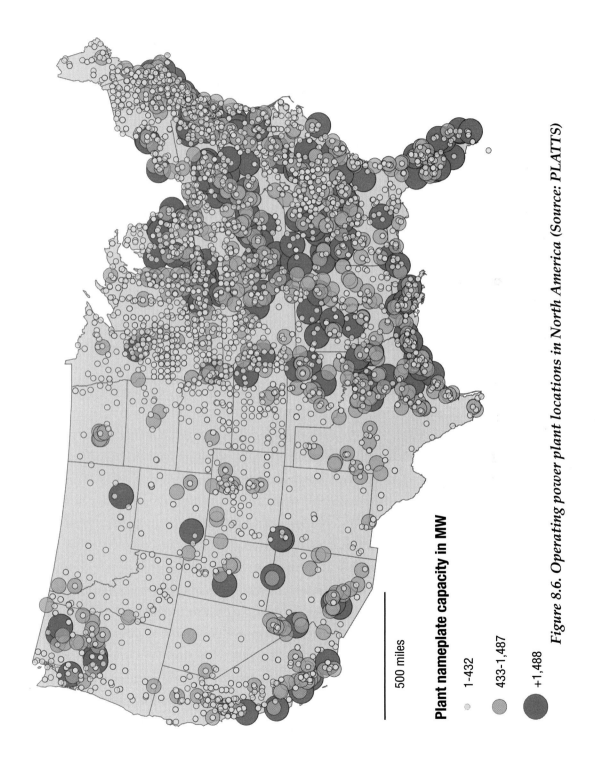

Plant nameplate capacity in MW

1-432

433-1,487

+1,488

500 miles

Figure 8.6. Operating power plant locations in North America (Source: PLATTS)

Occasionally, however, extraordinary conditions conspire to place demands on the system that it cannot accommodate. An extreme cold snap, accompanied by failure of a major electricity gridline, excessive solar or geomagnetic activity or other mechanisms can cause major disruption or even complete breakdown of the system. Such an event, or series of events, can be catastrophic.

Hospitals, airports, and certain other critical facilities, typically have independent power supplies of one sort or another. These are designed to prevent loss of critical function during a conventional, but uncommon, power outage from the grid. Offices, businesses, homes and public transport systems in major cities have no such independent power provision. Loss of power can instantly paralyze a city. Street lamps, subway trains, traffic lights, computers and mobile phone infrastructure can each suffer immediately from sudden loss of power. All of them can wink out simultaneously and plunge an area temporarily back into a primordial darkness where almost anything can happen. In a major city during, say, the start of the evening rush hour would not be a good time or place for such an event to occur, you might think – but such events have happened on a number of occasions and in several different countries. Expert predictions are that they will probably get worse unless significant investment is made to improve transmission and distribution networks.

Blackouts

The afternoon of August 14, 2003 was not particularly remarkable. It was a typical warm summer's day across large parts of the American Midwest and power consumption was high. At 2.02 pm, however, some high tension lines in northern Ohio, sagging slightly in the warm afternoon and more so because of the heavier than usual current flow, came in contact with some overgrown trees. The short circuit caused them to shut down. Under normal circumstances, this event would have been noted in the power utility company's control room, where human controllers monitor activity on the local power grid and work with their counterparts in neighboring states to ensure smooth flow of power around such trouble spots. Unfortunately, on that particular day, the alarm system failed to operate, leaving controllers oblivious to the problem. As chance would have it, at the same time, other controllers scattered across Ohio, Michigan, the Northeastern US and Southeastern Canada were routing large amounts of power hundreds of miles through the grid in response to ever-shifting demands. They, too, were unaware of the initial problem and power lines near the site, already struggling with the increased load due to current flow around it, were forced to carry ever higher amounts of electricity.

Over the next two hours a total of 22 high tension lines in Ohio failed while baffled controllers tried to understand what was happening. By 4.09 PM, only 3 minutes after the last of these lines failed, the Ohio grid was pulling 2GW of power from Michigan and a minute later, in response to this current surge, numerous other overtaxed power lines first in Michigan and then Ohio failed. Several power stations shut down, upsetting the system's already unstable equilibrium. The increased demand on other plants set huge currents surging through the system, one moment flowing eastwards, and literally the next second, westward, initiating a chain-reaction that would see more lines fail and one power station after another drop off the grid. The disturbance propagated faster than controllers could track it with their antiquated monitoring equipment and much faster than they could hope to intervene to try to stop it. By 4:13 pm, only seven minutes after the cascade began, 256 power stations were off-line, leaving 50 million people in the Northeastern US and in two Canadian Provinces without electricity.[19]

Fortunately, most of the power station shutdowns happened automatically, protecting generating equipment that human controllers would have been unable to save, since they had no idea how to control the wildly fluctuating grid.

At the time, it was the largest blackout in history but that dubious honor was not to last for long. Within two months, major blackouts had occurred in the UK (affecting 500,000 people), Denmark and Sweden and finally culminated in the biggest blackout ever on September 27, 2003, in Italy. The latter event came as a result of a power transmission line failure from Switzerland into Italy that subsequently caused overloading and failure of two 400 KV lines from France to Italy and left an estimated 57 million Italians without power for several hours and in some areas for as long as two days.[20]

Power outages (or blackouts, as they are commonly called) have happened many times, over the years. They are relatively uncommon but when they do occur, they cause major disruption and can wreak economic havoc. The famous Northeastern Blackout described above was estimated to have cost in the region of $6 billion in terms of lost productivity, etc. An historical survey of blackouts suggests that they are becoming more common and several analysts suggest that this is due to ageing infrastructure and control systems on commercial power grids. While electricity consumption has grown dramatically (2.1% per annum since 1989 in the US), the investment in transmission capacity has lagged (only 0.8% per annum over the same timeframe). However, the problem is not isolated only to the transmission system.

Generating capacity has languished due to permitting issues and several older plants have been shut down. Increased resistance to nuclear power during the 1990s either accelerated closure or deferred construction of several plants further reducing spare capacity and reducing safety margins. As pointed out in the Figure 8.7, below, the US alone may need to build as many as 2,000 new power plants in the period between 2000 and 2020 in order to meet demand and to replace retiring plants.

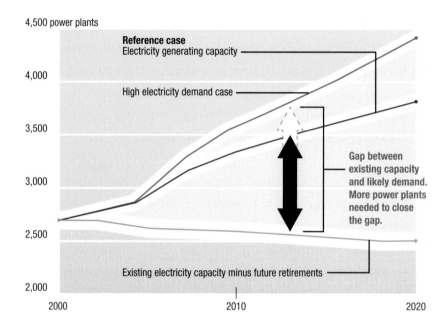

Figure 8.7. The US Needs More Power Plants (Source: EIA)

Storage

One of the biggest problems faced by electricity producers is the difficulty in storing the electrical energy. There are few practical ways to store electricity in commercial quantities given the enormous amounts of energy involved. Batteries can be used to store small amounts, thereby providing some protection for computers and other sensitive equipment, but the amounts of energy stored are trivial. Such uninterruptible power supplies are sufficient to provide enough energy for only a few hours of operation. The only practical method to store electricity, therefore, is to convert it into some other form of energy that can, in turn, be converted back to electricity when required. Such techniques are used by a number of hydroelectric power companies who buy excess capacity from the grid, during time of low demand. This "off peak" electricity is used to pump water from rivers, or from behind storage dams, into other holding facilities, at higher elevation and behind the principal hydroelectric generating dam. As demand increases again during peak hours, this water can be run back through the hydroelectric turbines to generate electricity again. This electricity can still be sold back to the grid at a profit due to differences in tariffs between "off peak" and "peak" times. Such pumped storage projects are relatively rare since they require suitable geography and topography and also because they consume more electricity than they actually generate. However, since their principal function is to store excess electricity generated at other facilities, allowing conventional base load plants to operate at peak efficiency, this is an acceptable trade-off. Worldwide pumped storage capacity is about 90 GW or about 3% of global generating capacity.[21]

An alternative system to store electricity involves the use of compressed air rather than water but the principle is the same. Off-peak electricity is purchased from the grid and used to drive compressors that pump air into large sealed underground reservoirs. During peak electricity demand, the compressed air is allowed to escape through turbines and the electricity generated is sold at a profit. Unfortunately, the process is not particularly efficient and only two commercial compressed air energy storage plants are in operation today, one in Germany and one in the USA.

Access

As noted in the introduction to this chapter, access to electricity by a nation's citizens is a key metric in assessing that country's state of development and in projecting its potential for future economic growth and increased demand. Throughout most OECD countries, access to electricity is almost 100 percent. However, in much of the developing world and transition economies, access is much lower and is often confined to cities and larger towns. As Table 8.1 below shows, an estimated 1.6 billion people around the world have no access to electricity and most of these people are located in sub-Saharan Africa, South Asia and China.

In order to meet the Millenium Development Goals, the total number of people without access to electricity needs to fall to below one billion by 2015. Looking at historical trends, we can assume that this represents an optimistic figure but we can also assume that access for a large number of these people will indeed improve over the next 10 years. We can also assume that consumption in the developing world, by those who already have access, will increase. In fact today, the disparity between the

	Population	Urban Population	Population without Electricity	Population with Electricity	Electrification Rate	Urban Electrification Rate	Rural Electrification Rate
	million	*million*	*million*	*million*	*%*	*%*	*%*
Africa	891	343	554	337	37.8	67.9	19.0
North Africa	153	82	7	146	95.5	98.7	91.8
Sub-Saharan Africa	738	261	547	191	25.9	58.3	8.0
Developing Asia	3,418	1,063	930	2,488	72.8	86.4	65.1
China and East Asia	1,951	772	224	1,728	88.5	94.9	84.0
South Asia	1,467	291	706	760	51.8	69.7	44.7
Latin America	449	338	45	404	90.0	98.0	65.6
Middle East	186	121	41	145	78.1	86.7	61.8
Developing Countries	4,943	1,866	1,569	3,374	68.3	85.2	56.4
Transition Economies and OECD	1,510	1,090	8	1,501	99.5	100.0	98.1
World	6,452	2,956	1,577	4,875	75.6	90.4	61.7

Table 8.1. Electricity Access in 2005: Regional Aggregates (Source: EIA)

electricity consumption by citizens in OECD countries and those in the developing world is quite stark. On a per capita basis, the average OECD citizen consumes 5-20 times more electricity than a citizen of the developing world. If we make that comparison with the average American, the disparity is even more extreme. Much of this is the result of higher levels of wealth and disposable income in the OECD as well as the availability and use of a wider array of gadgets and convenience devices by consumers in these countries. So, while there is an expectation that total electricity demand will increase significantly in the OECD countries, the increase, both in relative and absolute terms, will be overshadowed by the projected increases in the developing world. The latter will be driven both by increased access and increasing wealth, at least partially, resulting from the enhanced economic activity engendered by widespread electrification.

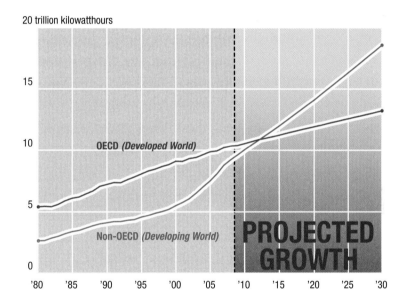

Figure 8.8. World Electric Power Generation by Region, 1980-2030 (Source: EIA)

For these reasons, growth in global electric power generation (Figure 8.8) and consumption will be rather lop-sided with more than 75% of that growth being centered in the developing world. Indeed, electricity demand in the non-OECD countries is expected to grow by 3.5% per annum over the next 20 years compared to only 1.3% per annum in the OECD. Most of the non-OECD growth will be in Asia, primarily in China and India, giving an even more skewed landscape to the projections, with demand growth in Asia topping 4.2% per annum. This amounts to a tripling of the region's electricity demand and gives some perspective to the fact that China is adding two 500 MW coal-fired power stations per week to meet rising demand. A similar, if somewhat less dramatic, increase in relative terms is seen in the non-OECD countries of (mainly, Eastern) Europe and Eurasia where demand is projected to double over the next 25 years. In established economies and in areas of low population growth, like the OECD countries of Europe, the increase in demand is expected to be much lower, only around 0.8% per annum amounting to a cumulative increase of a mere 25% over the next 25 years. However, even at these relatively low growth rates, such increases are still significant, given their higher absolute starting values. In fact, these increases would be higher still, were it not for the substantial improvements in energy efficiency attained over the past forty years or so.

Future Investment in Power Generation

Huge capital investment through 2035 will be needed if the world is to meet the projected increases in electricity demand. Indeed, the power sector will dominate the investment landscape of the energy business for the foreseeable future. The IEA estimates that the investment required amounts to some $11.3 trillion (Figure 8.9) and slightly more than half of that will be spent in the developing world.

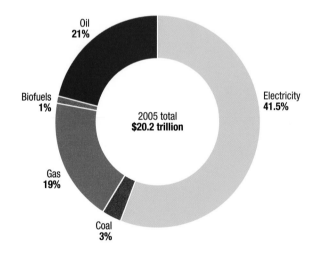

Figure 8.9. Capital Investment Required – Energy Sectors (Source: EIA)

This represents annual investment of about $500 billion over the next 25 years or about 1% of global GNP. Power generation equipment and facilities will account for about $5 trillion of the required investment while transmission/distribution infrastructure will account for the rest. Most of this investment will be in Asia and, unsurprisingly, given its strong economy and the percentage of its population who have not had access to electricity historically, the lion's share will be in China.

The existing transmission and distribution infrastructure in the developed countries is also showing signs of wear and tear and more than $2 trillion of investment will be needed simply to upgrade that infrastructure. A similar figure will need to be invested in new power generation capability in these same countries since a third of existing generating capacity is expected to be retired over the next thirty years.

Cleaning Up the Power Business

Despite the "cleanliness" of electricity at the level of the consumer, its generation does not have the best environmental credentials. Coal is the most widely used fuel for generating electricity (50% in the USA, 70% in India and 80% in China) and, therefore, it produces lots of pollution including sulfur dioxide and oxides of nitrogen (which together are responsible for so-called "acid rain"), mercury and certain particulates, detrimental to health (as mentioned in Chapter 7). Coal is also responsible for around 40% of the 29 billion tons of annual CO_2 emissions from energy use and the main culprit for the move to establish a carbon-constrained world.

Compounding these problems is the fact that most power plants are not terribly efficient. Only about 40% of the intrinsic energy of the fuel used in thermal generation ends up as electricity. More recently, however, so-called Combined Cycle generators, which extract additional energy from hot waste gases prior to venting them, have become more common. Such plants can raise efficiency to values of 55%, significantly reducing energy intensity and simultaneously reducing environmental fallout.

As a fuel for power generation, natural gas has seen its share increase dramatically in recent years and it is set to grow further partly driven by improved infrastructure to harvest and transport the product. Gas pipeline grids and the development of the global Liquefied Natural Gas (LNG) business have been instrumental in this growth and we are likely to see even more significant growth in the use of gas in the next decade or two. An attractive feature of natural gas is that it has far lower noxious byproducts than either coal or oil and virtually no ash. In addition, it produces significantly less CO_2 when it is burned. The sudden increase in demand for gas has caused shortfalls and price volatility in recent years and caused many power companies to switch back to coal. However, the new-found ability to access vast potential reserves of unconventional gas in very low permeability rocks (like shale) has dramatically changed the landscape for natural gas utilization. We are, therefore, likely to see it used increasingly for power generation applications.

Hydroelectricity, while appearing to be an ideal, renewable source of power, is also not without its problems. The habitat disruptions involved in dam construction and in flooding vast areas of land are considerable and many people may also lose their homes in the process. The flooding also drowns vegetation, which then goes on to decompose, producing CO_2 as if it had been burned. The impact of changed water flow patterns on wildlife and on agriculture may also be significant, reducing soil fertility and crop yields, disrupting migration routes, etc. Finally, of course, constructing hydroelectric dams and facilities depends on favorable geography and topography – and this combination of factors is only available in a limited number of locations, worldwide, restricting more widespread implementation.

When Intentions and Reality Collide

It is always the things that people do not think about or think through that control the results of energy experiments or expectations. It is also axiomatic that nothing is ever perfect in this business and things have a dark side no matter how bright they may appear. Consider what happened to China in 2009 and 2010. A severe drought that has plagued southwest China since August 2009 caused a painful shortage of water in an area where it had never been a problem in the past. But an ancillary villain from the situation was hydroelectric power generation which was partially blamed for the worsening water situation and at the same time was itself a victim of the drought.

China's hydroelectric power capacity, under the halo of low carbon energy and pushed by high profits aided by government subsidies, had been developed at a fast pace. In 2009 alone, 20 GW of new capacity was added and the national total hydroelectric capacity reached 197 GW, comprising 22.5 percent of the total national power capacity of 874 GW. The National Energy Commission had been considering increasing the total hydroelectric power capacity to 300 GW by 2020. The water-abundant and mountainous southwest China, including Yunnan, Guizhou and Guangxi, Sichuan Provinces and Chongqing District, had always been the preferred site for hydroelectric power. By the end of 2009 hydroelectric power capacity in this area amounted to 44.7 percent of the national total.

In August 2009, the hydroelectricity generated from this area was 30,640 GW-hr. In February of 2010, the generation was reduced to 9,540 GW-hr, a 68.9 percent decrease. Although affected by the season, this kind of decrease was unprecedented. Overall, hydroelectricity from the southwest power plants had been reduced by 40 percent due to the drought.

Under such circumstances, all coal-fired power plants have been in full-blast operation to keep the lights on. Since the capacity of coal-fired power plants is very limited in the area, power supply was in dire shortage with outages occurring frequently.

Guangxi has over 5,000 surface water reservoirs which are intended to prevent flood and remedy drought for agriculture. However, Guangxi had 2,284 hydroelectric power plants by 2008, which means that there has been one power plant for every two reservoirs. More power plants have been built since then. As soon as a power plant is built, water has to be discharged to generate electricity, no matter whether the downstream needs water or not for agriculture or human consumption. After the drought lasted for more than 6 months, the water level in many reservoirs had been dangerously low but the power plants still kept discharging water for power generation. Some reservoirs became dead water pools or dried out altogether. As a result, about 22.7 million people haven't had enough drinking water, according to March 23, 2010 data. The local people complained that if the discharge of water from the reservoirs had not been following the needs of the power plants, the ill effect of the drought might not be so severe.

The drought increased the use of coal by 4 to 5% for power generation. The ill effects of the drought may force China to think the next time the country embarks in a new push for even more hydroelectric power plants.

Nuclear Power Generation

By an odd twist of fate, that old *bete noire* of environmentalists, nuclear power may, actually, prove to be the best bet in the intermediate term, for providing relatively cheap and carbon-neutral electricity. Reviled for years, with many politicians responding to rising public concerns, orchestrated by environmental pressure groups and the media, several governments have been intent on decommissioning their nuclear power facilities. However, nuclear energy is a key part of many countries' power generating portfolio supplying 78% of France's electricity, 48% of Ukraine's, 26% of Japan's and 20% of both the USA and UK's. In 2010 there were 439 nuclear reactors operating in 30 countries around the world with a generating capacity of approximately 375 GWe (Figure 8.10) – supplying 14% of the world's electricity.

While China had only 11 commercial nuclear reactors in operation in 2010, at least 23 more were under construction and 34 more were already planned. The total number of reactors under consideration for the country by 2030 is about 180 with a capacity of as much as 200 GWe. India also plans to build additional reactors. In 2010 there were 19 commercial reactors in operation (plus 5 research reactors), 4 under construction and another 60 planned or proposed. In total, worldwide, 150 new nuclear reactors were planned and over 350 more reactors were proposed.[22] During the 1980s, a total of 218 reactors went on line or, on average, one start-up every 17 days. With the growing demand for electricity, particularly in Asia, it is currently considered perfectly conceivable that, over a ten year period from 2015, one new reactor will start up every 5 days. (See Appendix F for information on planned and currently commercial nuclear power plants)

Public fears over nuclear power generation tend to focus on issues like radiation and the risk of accidents, leakage or terrorism. Several high profile nuclear accidents, including those at Three Mile Island in Pennsylvania, USA and Chernobyl in the Ukraine in the former Soviet Union, have provided graphic evidence of the potential dangers of nuclear power. The Three Mile Island event in 1979 was largely contained and resulted in no direct fatalities, despite a partial meltdown of the nuclear reactor core. The Chernobyl incident, on the other hand, caused the deaths of 56 people from radiation exposure and forced a mass evacuation and resettlement of 336,000 people. The WHO has estimated that the Chernobyl accident contributed to an additional 4,000 deaths due to the longer term effects of exposure to the high levels of radioactivity released from the plant. Almost 25 years after the incident, a 30 kilometers exclusion zone around the reactor is still in force and will probably remain so for the next century, at least. The remains of the reactor itself will not be safe for 20,000 years.

Such contamination serves only to underline the fact that all reactors generate long-lived radioactive waste that must be stored in special facilities, in perpetuity. This causes concerns among the general public, as does the production of even more dangerous fissile material (e.g., plutonium) that could be used to make nuclear weapons, by some types of so-called breeder reactor. However, despite these concerns and misgivings, most governments are resigned to a renaissance in nuclear power generation to meet GHG emission targets and reduce reliance on fossil fuels.

Nuclear energy is not the only carbon-neutral power source, of course. There are at least three other sources which are not really new but have been mentioned and even developed with varying degrees of vigor for 40 years. These include wind, solar and geothermal.

Wind generated electrical power, produced by the flow of air currents over giant, modern-day windmills (Figure 8.11), is the most important of the three today.

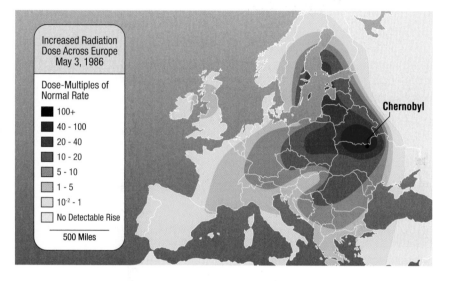

Figure 8.10. The Chernobyl disaster was the worst nuclear accident to date. Meltdown of the reactor core caused a steam explosion and fire which sent a contamination plume across Europe in April 1986. (Source: Open University)

Figure 8.11. Modern-day windmills utilized for wind generated electric power

Countries like the USA, Germany, Spain, China and India lead the world in installed capacity of wind turbines, largely as a result of generous tax credits. Table 8.3 shows the top countries in terms of installed capacity and Figure 8.12 shows the cumulative growth of wind energy market, globally. Clearly, the business has experienced exponential growth over the past 10 years or so and, if anything, that growth is accelerating. With the passing of legislation to encourage renewable energy ventures, China more than doubled its wind power capacity in 2008 and has plans to almost double that again by 2010. Forecasts put China's wind generated power capacity at 50,000 MW by 2015. However, the strongest growth in this sector over the past year, in absolute terms of new installed capacity, was in the USA which added over 8,000 MW in 2008, overtaking Germany as the country with the largest wind generation capacity.

Table 8.2. Installed Wind Turbine Capacities – Top 10 Countries (Source: GWEC)

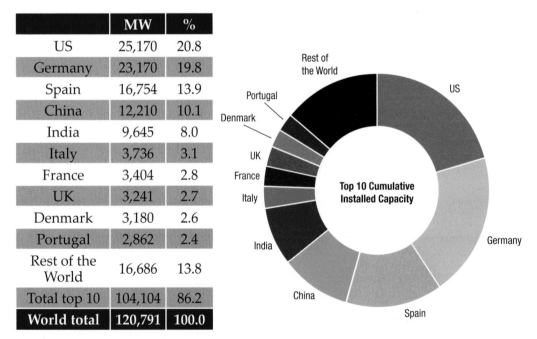

	MW	%
US	25,170	20.8
Germany	23,170	19.8
Spain	16,754	13.9
China	12,210	10.1
India	9,645	8.0
Italy	3,736	3.1
France	3,404	2.8
UK	3,241	2.7
Denmark	3,180	2.6
Portugal	2,862	2.4
Rest of the World	16,686	13.8
Total top 10	104,104	86.2
World total	120,791	100.0

These figures may look impressive and there is no doubt that wind power is becoming mainstream but it should be noted that actual electricity generated is 25-30% of the installed capacity in good locations and a lot less elsewhere, due to the variable nature of the wind. This variability also means that wind-derived power cannot be used to support base load requirements, without significant and expensive changes to transmission and distribution infrastructure and without some mechanism to provide substantial redundancy, in anticipation of scenarios where the wind drops and power becomes unavailable. Wind power enthusiasts point out that the grid integrates power from many sources already and that it can easily deal with the fluctuations from such a minor contributor as wind. This may, indeed, be true at current levels where wind contributes barely 1% of grid power. However, with ambitious plans (and perhaps highly unrealistic targets) for wind power to provide as much as 20% of power demand by 2030, the need for redundancy of one form or another will certainly increase, too.

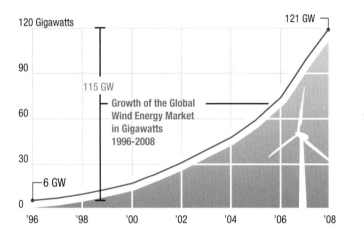

Figure 8.12. Growth of the Global Wind Energy Market (Source: GWEC)

As noted previously, because of large government subsidies to support development of wind technology, the efficiency of wind turbines has improved dramatically and some proponents argue that it will soon be cost competitive with coal generation, without the detrimental pollution and potential climate effects. Clearly, it has come a long way in a short time and seems destined to provide an increasing fraction of renewable electrical power.

Solar energy has been harvested by mankind since the dawn of history, using the sun's heat to warm water, dry crops, and similar functions. Of course, all food grown on the planet is dependent on the sun's energy so we also harvest this energy simply by eating and one could even assert that other energy sources like wind, hydroelectric, and even coal, oil and gas, are all ultimately derived from the sun's energy. These involve indirect uses of solar energy, however, rather than the direct use of the incoming energy arriving each day on the earth's surface, which ranges from 125-375 W per m² in the United States. An interesting fact, although not relevant in this discussion, is that each hour, the incoming (incipient) solar radiation striking the earth is greater than the human race's energy consumption for an entire year. The ability to efficiently convert that energy directly into power is the holy grail of the solar energy business. Clearly the process is not only a function of efficiency but also a strong function of the areal extent of solar collectors required for a meaningful solar power generation. For example, the incoming radiation figures quoted above are enough to light one small light bulb per square meter of solar collector. Figure 8.13 maps the average incoming solar radiation globally – averaged over a period of 3 years.

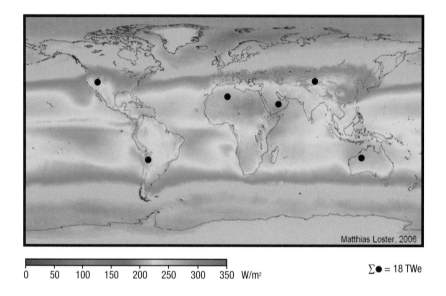

Figure 8.13. Average Incoming Solar Radiation (Source: Matthias Loster)

The actual task of fabricating, installing and maintaining a colossal array of required solar panels, then transmitting and delivering the power to consumers around the globe, is likely to ensure that this remains no more than an idle fantasy for the foreseeable future.

In the context of the power industry, solar energy is currently a niche technology, providing only a tiny amount of the world's energy requirements. However, again buoyed by massive government subsidies, its use is increasing dramatically – 40% per year from 2005 to 2010, admittedly from a very low base.

There are three main types of direct solar energy – hot water, solar thermal and photovoltaic. The first of these is primarily of interest not for generating electricity but rather for reducing electricity demand by providing free hot water and, perhaps, space heating. Simple systems for the average home are claimed to reduce greenhouse gas emissions by about 4.5 tons per annum per installation. Solar thermal furnaces, on the other hand, are intended to be used for commercial scale electrical power generation and focus the sun's energy using parabolic mirrors or arrays of focused light collectors (heliostats) to heat thermal oil or, in the latter case, liquid sodium metal, to very high temperatures. The oil or liquid sodium are, in turn, passed through heat exchangers, heating water to produce steam, which is used to drive turbines to generate electricity. The largest solar electric plant in the world, the 350MW SEGS, in California's Mojave Desert, uses parabolic trough mirrors as does a smaller 65MW plant in Nevada. No commercial systems based on the heliostat design are yet in use, although one is planned in South Africa.

Photovoltaic (PV) systems use a phenomenon first observed by Becquerel in 1839 to generate electricity directly by the effect of light on semiconductors (the photoelectric effect). The first photovoltaic cell, produced in 1883 using selenium, had an efficiency of only 1% and was certainly impractical for generating electricity, commercially. The modern age of solar power cells arose from the accidental discovery at Bell Laboratories in 1954 that silicon, a much more common element, doped with certain impurities became very sensitive to light. This observation led to the first practical PV cell, which had an efficiency of 6% and made it feasible to use in specialized applications like the then-nascent space industry.

Figure 8.14. Polycrystalline Silicon PV Array

With renewed interest, novel materials, improved manufacturing techniques (using processes developed for making microprocessor chips), support from government via subsidies and big investment from industry, efficiencies have improved in recent years. Conversion efficiencies of silicon-based systems have been reported to be as high as 25%, although commercial modules are normally around 15%. Some exotic semiconductors (gallium arsenide alloy) have achieved efficiencies approaching 40% in the laboratory but no commercial versions are yet available. Recently, researchers have devised alternative schemes to improve efficiency of PV cells, including concentrating the sun's rays using lenses and mirrors – and using high surface area nanoparticles to increase the capture ratio of the incident energy. Other schemes include configuring the cell in such a way that incoming light is reflected internally, traversing the thin film several times. Also, using dyes that are sensitive to infrared or ultraviolet radiation can enhance performance by ensuring that a greater portion of the sun's spectrum is transformed into electrical energy.

Despite these enormous improvements, solar PV today only provides a miniscule amount of global energy (0.039%). Yet, it is certainly the fastest growing segment of the energy business and one that is continuing to attract enormous investment. With legislation and subsidies designed to expand its use, its market penetration over the course of the next decade is bound to increase but still be considerably below 1% of power generation.

In order to provide usable power for domestic applications, output from PV arrays must first be converted to AC via an inverter. However, by providing relatively stable power continuously throughout the day in sunny locations, PV can reduce base load requirements significantly. Indeed, with sufficient areal coverage, PV can provide distributed power generating capacity and any excess can be sold back to the grid.

The final renewable power source with sufficient capacity to contribute to energy requirements is geothermal – the use of the Earth's natural internal heat. This represents a large source of potential energy but one that, like solar and wind, requires major investment and, more to the point, favorable locations near geothermal anomalies, places on earth where the mantle is near the surface, heating the ground and resident water which can be produced as either steam or liquid and emit energy for power generation, space heating, etc. There has also been some experimentation using hot dry rock where water is injected in wells and then produced in other wells in an attempt to mine heat. These ventures have not produced any commercial results.

In 1972, installed capacity for generation of electricity from geothermal sources worldwide was about 800 MW. Annual production of electricity from geothermal sources today is around 67,200 GWh from 10.7 GW of installed capacity in about 25 countries. The world's largest producer of electrical power from such sources is the United States, with 3.0 GW of capacity. Philippines, with almost 2 GW of installed capacity, is the second biggest producer of geothermal electricity, deriving over 20% of its annual electricity budget from such sources.[23] Other countries with significant geothermal power generation include Indonesia, Mexico, Italy and New Zealand. In fact, the latter two countries were the first to construct commercial geothermal power plants. The Italians led the field with their facility in Larderello which began producing electricity in 1911. Surprisingly, it was not until 1958 that the world's second plant came online in Wairakei in New Zealand's North Island.

Some very optimistic estimates of usable geothermal resources that could be developed within the next twenty years in the US are of the order of 13 GWe. This type of estimate was repeated in the 1970s but it did not materialize. From a global perspective, estimates of geothermal resources put the available capacity, using existing technology, in the range of 35-75 GWe.

New Trends and Evolving Infrastructure

As noted at the beginning of this chapter, transmission of electricity is made possible by the ability to "step-up" the voltage and, simultaneously, reduce the current using a transformer – then to reverse the process closer to the consumer. This minimizes losses caused by resistance in the transmission lines. This exercise is particularly simple with AC and was one of the main reasons for the selection of the latter as the system of choice in the early days of electrification. However, the system is not perfect. While transformers themselves are extremely efficient devices (typically 95-99.5% efficiency), the transmission lines themselves offer resistance and the intrinsic, fluctuating character of AC gives rise to a variety of losses (reactance, impedance, eddy currents, etc.). Typically, in developed countries, transmission losses amount to around 7%. Over very long distances, though, line losses with AC become more and more significant.

One solution to this problem is to increase the voltage to ever higher levels. This requires the use of higher and wider pylons to separate the high tension lines from the ground and from each other, as well as ever-greater insulation. Also, the use of such high voltages (currently the highest is around 1.2 million Volts) increases losses due to other phenomena like "coronal arcing."

An alternative approach to the problems implicit in long distance power transmission is to use high voltage DC (HVDC) rather than AC. This is made possible by improved solid state devices (thyristors) that were developed in the 1960s and were thus not available in the late nineteenth century when Edison and Tesla were locked in bitter rivalry over which system would gain dominance.

HVDC offers a number of potentially significant benefits over the more conventional AC transmission systems.[24] First of all, it requires much less in the way of transmission infrastructure since it can be carried using only two conductors on much simpler and smaller pylons. This has important implications on issues such as "rights of way" and the amount of land required for hosting the transmission line. HVDC pylons need not be as tall or as wide as those carrying AC since they are not susceptible to the same interactions between adjacent conductors or between conductors and earth as AC lines.

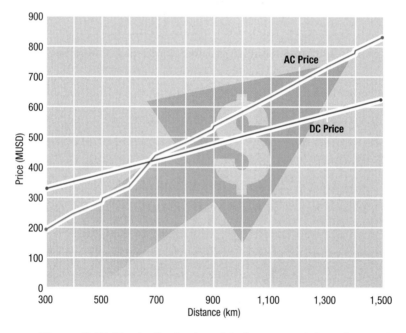

Figure 8.15. Typical relationship between AC and HVDC
infrastructure costs vs. distance (Source: Siemens)

For transmission over very long distances, HVDC losses are significantly lower than those of AC systems. Also, when it comes to delivering power between so-called "asynchronous" grids (i.e. AC grids operating at different frequencies, 50 Hertz vs. 60 Hertz, for example), HVDC offers an elegant solution since it can be converted to, or from, any desired frequency of AC, as required.

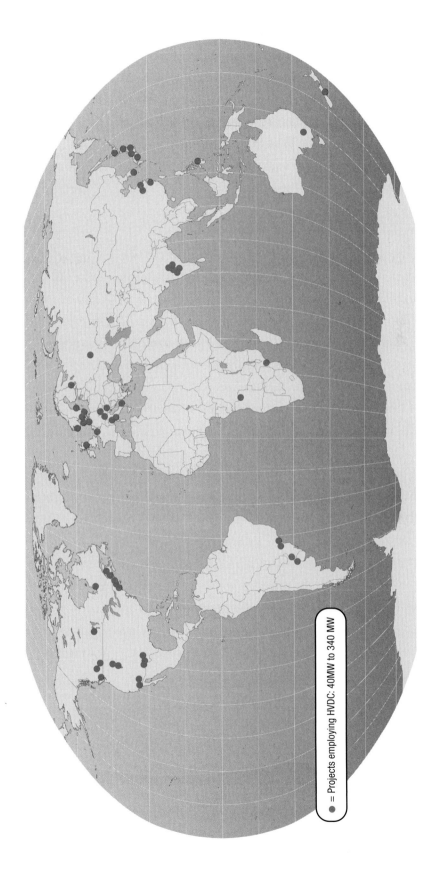

Figure 8.16. Some major electricity infrastructure projects

● = Projects employing HVDC: 40MW to 340 MW

In a world of increasingly interconnected transmission infrastructure, across national borders and even, perhaps, between continents – where vast amounts of power need to be carried for long distances – there is certainly a case to be made for HVDC. Many projects have already made use of this technology and many more are considering novel concepts for electricity generation, transmission and distribution as a result of the benefits already seen.

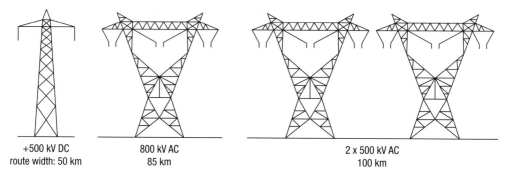

| +500 kV DC | 800 kV AC | 2 x 500 kV AC |
| route width: 50 km | 85 km | 100 km |

Figure 8.17. Three alternative pylon configurations for a 2,000 MW transmission system

One such idea is the siting of generating facilities adjacent to coal mines, far from population centers, where power can be generated cheaply and the cost of moving coal is minimal. Similar concepts could be applied to those countries that have vast hydroelectric potential but small local markets (e.g. Zambia, Norway). Such countries could benefit from the sale of electricity to countries many thousands of miles away, transmitted with low losses and cheap transmission infrastructure, courtesy of HVDC technology.

Ideas also abound for the use of HVDC to carry power from vast photovoltaic arrays in places like the Sahara to the electricity markets of continental Europe – or to use it for integrating the power from thousands of wind turbines to fill pumped storage reservoirs in Norway's hydroelectric dams. These could then be used to provide base load requirements for much of Europe, again carried on HVDC transmission lines.

Thus, a hundred years after this revolutionary form of energy became commercially available, the electricity business continues to evolve – as do the in *advance* investment opportunities – and technical improvements in the generation, transmission, distribution and end uses of electrical energy are still being researched, developed and implemented.

CHAPTER 9

Alternative Transportation Fuels

Chapter 9
Alternative Transportation Fuels

As pointed out in several places throughout this book, oil and gas will remain by far the dominant fuels for the world, for many decades to come. Regardless of the doomsayer predictions that we will run out of these finite commodities within the next few years, we are confident that this will not happen. However, there is no doubt that the price of these materials will increase with time, not just because of physical reasons but primarily because the western world has decided to divest itself from them, leaving the control to oil producing countries, many of them using oil and gas as a means for national emancipation and international relevance and power. However, such increases in oil and gas prices will not be relentless, they will be tempered by other economic events, such as the 2008-2010 economic crisis. No doubt we will see periods of soft pricing for these valuable, fungible commodities in the future, just as we have seen in the past. However, such price fluctuations, in response to global crises, political tensions, temporary imbalances in supply and demand or speculation by commodity traders, will only be superimposed upon a long-term, generally upward trend in oil and gas prices.

Higher prices usually act to curtail demand to some extent but they also act as a catalyst for increased research into energy efficient technologies and alternative sources of energy. As noted in Chapter 8, while the use of oil has been superseded, in areas like power generation, by the switch to gas, a return to coal or the stirrings of renewable energy supplies, its use as a transportation fuel has been virtually unassailable. In the US, almost 70% of the oil consumed is used for transport purposes, primarily motor gasoline, diesel, and jet fuel. On a global basis, approximately 30% of world primary energy supply is consumed by the transportation sector and, of that fraction, 96% of the fuel used is derived from hydrocarbons.

So, clearly, there is great interest in finding ways to reduce this consumption in the face of rising oil prices and uncertainties over future supply. In recent years several alternative fuels have begun to enter the market. Certainly, this has largely been the result of government subsidies.

However, price and energy security have not been the only reasons. The notion that it may be possible to reduce emissions of carbon dioxide, while simultaneously reducing the dependence on oil imported from a few countries, is certainly a seductive one. Using fuel products derived from plants has the potential to be carbon-neutral (not adding CO_2 to the atmosphere), since those same fuel products have been made by plants that recently removed the CO_2 from the atmosphere, in the first place, in the process of photosynthesis. Burning such fuels, therefore, while still generating CO_2 just like other carbon-based fuels, results in no net increase to current atmospheric levels of the gas.

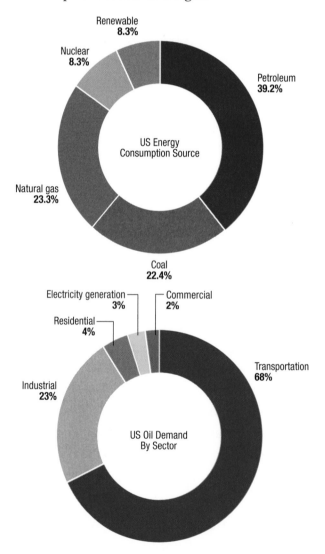

Figures 9.1., 9.2. Energy consumption by source and oil demand per sector of the economy. Most oil is consumed in the transportation sector. (Source: EIA)

Figure 9.3. Increasing public activism, fueled by concerns voiced by many scientists and encouraged by media and environmental groups, have forced policy makers to consider alternatives to conventional fuels.

Before discussing the various alternative fuels, it is important to define the potential markets for these fuels and to differentiate which amongst them are, at least at this stage, likely to be viable replacements for those already in use. Thus, in the world of transportation, the three main modes of transport, land, sea and air will almost certainly require different solutions. On land, the fuels in use are overwhelmingly gasoline and diesel (No.2 fuel oil). At sea, heavy fuel oil (No.6 oil) is used for most ships' engines while, in the air, both commercial and military aircraft use kerosene (Jet A and Jet A1). Current replacements for gasoline, diesel and fuel oil for use on land or sea are certainly not perfect but tremendous progress has been made in some areas due to new thinking and novel technology introduction. However, in spite of this progress, the problem is still a daunting one and the scale of the challenge to replace these vital commodities is enormous. In the US alone, gasoline consumption exceeds 378 million gallons per day or about 140 billion gallons per year.[25] Diesel fuel consumption is around 60 billion gallons per year. With current technology, out-and-out replacement of either of these fuels would be impossible, anytime soon. In the same way that oil is a finite resource, our ability to grow enough fuel crops will depend on the availability of finite land and water resources, among other things. For example, if all corn grown in the United States is converted to motor vehicle fuel it will be less than 20% of the gasoline demand. If all soybeans were converted to bio-diesel it would be less than 4% of the diesel demand.

The potential solutions for ground transportation are probably simpler and more viable than anything that is yet in the pipeline, as it were, as a carbon-neutral replacement for aviation kerosene. Despite some high-profile publicity exercises undertaken by several airlines and aircraft manufacturers to demonstrate their commitment to "sustainable" air travel, the practical hurdles for the aviation industry are considerable. There are numerous reasons for this. First, aircraft must carry their fuel aloft, so it is important that the fuel should have a high energy density. In other words, each unit mass of fuel must pack a lot of energy. The fuel must also be relatively easy and safe to transport and dispense, have a high flash point and remain usable at the very low temperatures encountered at high altitude. On all counts, hydrocarbons are excellent fuels exhibiting many of the desirable characteristics listed and several of the alternatives discussed in this chapter for ground-based applications simply cannot be considered for aircraft. Thus, commercial aircraft running on hydrogen or on batteries are simply out of the question. Some of the specific challenges for other potential aviation fuels are discussed below, as we examine the general state of alternative fuel technology today. Again, however, it is also worth reiterating the scale of the problem that we face. The aviation industry, globally, uses some 5 million barrels of aviation-grade kerosene (~200 million gallons) per day.[26] Even if they were usable, which is by no means obvious, growing sufficient biofuels to supply aviation alone would require planting 1.4 million kilometers^2 of oil-yielding crops, an area roughly twice the size of France.

The first of the "sustainable" or "renewable" transport fuels to appear in significant volume was ethanol, a simple alcohol produced, originally, by the fermentation of sugar by yeast. More recently, ethanol has been produced from maize (corn), a route that has been enjoying significant subsidies from the US Federal Government. The ethanol produced in this way is expensive and would not be viable without such subsidies. The long term hope, however, is that newer, more efficient ways to produce ethanol will be researched and commercialized, in response to the huge market created for the industrial product. In particular, the hope is to move from food starch sources (of which corn, or maize, is one) to waste cellulose sources, like corn hulls, lumber waste, grass, etc. Today, ethanol is blended, usually with gasoline as "Gasohol," where it acts as an "oxygenate," improving the burn efficiency and reducing emissions of pollutants. This practice dates originally from the 1930's when both the UK and Germany employed gasoline/ethanol blends to reduce their reliance on imported petroleum products. However, despite this focus on blended products, ethanol can also be used on its own, as a 100% replacement for gasoline. It was the original fuel considered for the Otto internal combustion engine and touted as an alternative fuel for Henry Ford's Model-T automobile.

Figure 9.4. Sir Richard Branson displays the source of the biofuel used on the first flight by a commercial aircraft. One engine operated on a blend of conventional jet fuel blended with 5% biofuel made from coconut and babassu nut oil. The trial required 150,000 coconuts to produce oil for a little over a ton of fuel (22 tons of blend).

The resurgence of ethanol as a transportation fuel can be traced to 1975 when, in response to increasing oil prices and rampant domestic inflation, Brazil embarked on an ambitious program to develop ethanol as an alternative to petroleum-based fuel products. The ethanol was produced by fermentation of sugar extracted from sugar cane, a commodity that had lost much of its commercial value, at the time, due to competition from sugar beet. The Pro-Alcool program was a government-funded initiative intended to phase out petroleum-derived fuel products and replace them with domestic ethanol, thereby improving the country's economic situation and its energy security. As a result of the program, from 1982 to 1988 more than 90% of Brazil's cars were being run on pure ethanol or on gasohol, the popular name given to blends of gasoline and ethanol.

Figures 9.5 and 9.6. In Brazil, sugar cane is extensively planted and transformed into ethanol in factories close to the plantations. Harvesting the cane is still back-breaking work.

Successful though the program was, however, it collapsed in 1990 due to a sudden increase in world sugar prices and falling oil prices, a combination that conspired to create a shortage of ethanol and reduced petroleum fuel products prices to levels where they were cheaper than ethanol. Despite this setback, however, that early Brazilian experiment made it, for a time, the world's largest, and, still today, the lowest cost, producer of ethanol, with 2009 production of around 6.58 billion gallons. The cost to produce a gallon of ethanol in Brazil is at least 30% percent less than US (corn-derived) ethanol and about 60% less than in the European Union. Since 2000, increased oil prices coupled with the introduction of so-called Flex-fuel vehicles that can run on either pure ethanol or gasohol, have resulted in significant renewed interest in ethanol as a transport fuel.

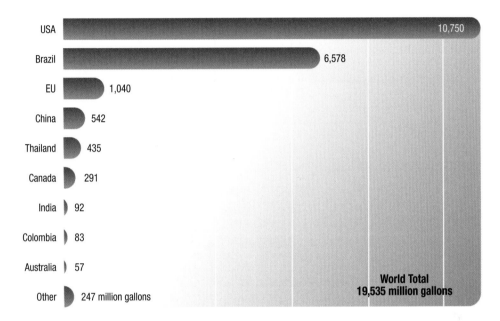

Figure 9.7. Ethanol production for 2008. (Source: Renewable Fuels Association)

Until very recently, however, Brazil was still largely an anomaly and very few countries had attempted to emulate its success with ethanol. Part of this was due to issues, both perceived and real, relating to the use of ethanol as a fuel. Corrosion of engine components, absorption of water (by pure ethanol), and reduced distance per tank of fuel (due to the lower energy density of ethanol, mentioned above) were among the concerns of motorists. The rising price of oil and the popularly perceived benefits of ethanol as a renewable "biofuel" with much less impact on the environment have assuaged these and other concerns, at least temporarily. So, too, has the routine inclusion of alcohol in fuel blends available at the pump, replacing older oxygenates like MTBE. In the US, gasohol blends containing 15% alcohol are now the norm in several states and, for many years, Sweden has employed a blend containing 5% ethanol.

The US government has played a key role in promoting ethanol in the past few years, providing huge incentives to US farmers to grow maize for ultimate conversion to ethanol. From a production of only 6.2 billion liters (~1.6 billion gallons) in 2000, the US overtook Brazil as the world's biggest producer of ethanol in 2006, eventually producing nearly 40 billion liters (~10.75 billion gallons) of ethanol in 2009. While initially widely hailed as a positive program, helping to reduce America's reliance on foreign oil, critics of the program are scathing about its ineffectiveness on almost all fronts. These criticisms are discussed, in more detail, below.

Figures 9.8 and 9.9. The US government has provided subsidies for ethanol produced from corn

Ethanol, however, is not the only biofuel and, indeed, is probably not the most attractive. Other possible candidate molecules have been proposed and, in some cases, experimental or commercial prototype plants have been set-up to demonstrate proof of concept.

Butanol, an alcohol similar to ethanol but with four carbon atoms instead of just two, is one such alternative fuel. It can be synthesized chemically using conventional technology or can be produced from carbohydrate and cellulosic feedstock by bacteria and specially adapted yeast strains. When produced by fermentation, in this way, it is often referred to as Biobutanol, to signal its "green" credentials. From an energy density perspective, butanol is a much more attractive fuel than ethanol, requiring only about 10% more volume than gasoline to travel the same distance (compared to a 40% increase for pure ethanol). It can, allegedly, be run "straight" in existing gasoline engines, with minor modifications, or can be blended with gasoline, just like ethanol. Butanol is also more hydrophobic (i.e., it mixes less readily with water) than ethanol, due to its longer carbon chain, making it more tolerant to water contamination, more miscible with gasoline and less likely to separate on storage. Finally, butanol has a relatively high flash point, making it safer to handle and store than either gasoline or ethanol.[27]

Despite these significant advantages, very little has been done to date to commercialize biobutanol and it is seldom referred to by auto manufacturers. One problem that has hindered biobutanol's promise as a fuel is its toxicity to the bacteria and yeast strains that produce it during the fermentation process. While some yeasts are capable of surviving in fermentation mixtures containing up to 15% ethanol, they are killed at butanol concentrations of only 2-3%. This makes it costly and inefficient to separate butanol by distillation and is likely to hinder growth in the use of this promising biofuel. However, the British oil company, BP, and the chemical giant, DuPont, currently have a joint venture investigating improved methods for production and new uses for biobutanol.[28]

Table 9.1. Properties of common fuels

Fuel	Energy Density	Air-Fuel Ratio	Specific Energy	Heat of Vaporization	RON	MON
Gasoline	32 MJ/L	14.6	2.9 MJ/kg air	0.36 MJ/kg	91–99	81–89
Butanol	29.2 MJ/L	11.2	3.2 MJ/kg air	0.43 MJ/kg	96	78
Ethanol	19.6 MJ/L	9.0	3.0 MJ/kg air	0.92 MJ/kg	129	102
Methanol	16 MJ/L	6.5	3.1 MJ/kg air	1.2 MJ/kg	136	104

Another interesting molecule with many desirable properties as a biofuel is 2,5-dimethylfuran (DMF), a recent entrant into the contest. A novel synthetic method, pioneered at the University of Wisconsin, has suddenly made DMF an attractive potential fuel.[29] The compound is stable, has a low freezing point, high boiling point and has a high energy density, some 40% higher than ethanol. It is produced catalytically from fructose in a two-step process, making its manufacture from raw materials much faster, and more amenable to process optimization by chemical engineers, than the fermentation process used to make ethanol. As noted elsewhere, some 67% of the energy required to make ethanol is consumed in the fermentation and distillation process. In the manufacture of DMF, however, the product separates spontaneously from the aqueous (water) phase, significantly reducing energy requirements, since no distillation is needed. Thus, DMF promises to deliver twice as much energy as ethanol, per unit of energy consumed in its manufacture.

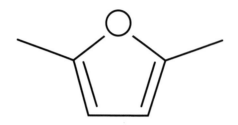

Figure 9.10. Dimethylfuran (DMF) is a recent contender as a so-called second generation biofuel. It has a high energy density, can be made catalytically from sugar and separates from the reaction mix without distillation, saving energy. It is also miscible with existing hydrocarbon fuels.

Apart from gasoline, the other major hydrocarbon fuel used for ground transportation is diesel, traditionally a mixture of higher alkanes, cycloalkanes and usually some aromatics. Increasingly, such mineral diesel is called petrodiesel, to differentiate it from fuels derived from renewable sources. With its lower volatility and longer carbon chain length, diesel fuel is used in high-compression engines that require no spark to initiate ignition. The original patents for this type of engine featured the use of low volatility fuels that would vaporize and auto-ignite when injected into the engine cylinder filled with air heated to several hundred degrees Centigrade by rapid compression. Many fuels were suitable for such an engine and, as an historical footnote, the Otto company built a compression-ignition engine that ran on peanut oil (arachis oil) for the Paris World Fair in 1900, on the request of the French government. Later in life, Rudolf Diesel, whose name is usually associated with the compression-ignition engine, promoted the use of plant-derived fuels in a talk given to the Institute of Mechanical Engineers in London.[30] It is not too surprising, therefore, that oils derived from various plant sources should, once again, be considered as potential substitutes for petrodiesel, giving rise to the general term, "biodiesel," a name that first surfaced in a 1988 technical paper from China.

In recent years, "biodiesel" products, derived from a variety of natural plant and vegetable oils have appeared and are already in use in several places, replacing some petrodiesel, typically in "captive fleet" vehicles used by government or municipal authorities, e.g., buses, garbage trucks, postal vehicles, etc.[31] Again, in the absence of subsidies, it is not terribly efficient or cost-effective today to use vegetable oils to produce biodiesel but improvements in the process technology and, perhaps, the types of oil are expected to help.

Figure 9.11. Jatropha, a hardy shrub that produces oil-rich seeds for non-food use is receiving much attention as a potential crop for biodiesel production. Jatropha is resistant to drought and grows on relatively poor quality land that is unsuitable for food production yielding a clean, high quality oil.

The preferred oils for use in biodiesel include rapeseed, corn oil, soybean oil, palm oil and jatropha as well as reclaimed waste (i.e. previously used) food oils from commercial enterprises, like restaurants and snack food manufacturers. The utility of each particular oil varies, depending on the oil type, source and chemical composition. Some are suitable for use "as is," without chemical modification, as direct substitutes for petrodiesel or blended with the latter at typical ratios of 2-20%. Those that can be used in this way, however, (so called SVO – Straight Vegetable Oil) are the exception. The majority of vegetable oils require chemical treatment in a process called "transesterification," where they are reacted with an alcohol (usually methanol, which is cheapest) in the presence of a caustic soda catalyst to yield the methyl ester of the oil in question and it is this product that is typically referred to as "biodiesel." The process mentioned above is simple and can be carried out with rather basic equipment and minimal training, making it ideal in countries with limited infrastructure or lacking sophisticated chemical processing facilities. A newer technique, pioneered by the Finnish oil company Neste, involves direct hydrogenation of vegetable oils to produce an oxygen-free biodiesel of very high purity. The benefit of this approach is that it reduces waste, eliminating the production of glycerol as a by-product, thereby improving efficiency and increasing final product yield. The biodiesel produced in this way also has a higher energy density since it contains no oxygen, unlike the more conventional trans-esterified biodiesels.[32]

Regardless of the exact production method, biodiesel exhibits excellent fuel characteristics for both for compression-ignition engines and turbines (as used on jet aircraft) with performance that often exceeds the technical specifications for petrodiesel. It burns cleaner with little sign of soot and particulates, reduced carbon dioxide and carbon monoxide emissions, zero sulfur dioxide (since vegetable oil contains no sulfur) but some increase in nitrogen oxides. Also, because it has better lubricant and solvent characteristics than petrodiesel, biodiesel reduces the need to incorporate extra additives in the fuel. While these fuels can be used in jet engines, they are unsuitable as aviation fuels unless blended with conventional kerosene. The main reason for this is the need for commercial jet fuels to remain liquid at the very low temperatures experienced in flight (-40 – 47 °C). All of the commercial airline trials using biofuels, to date, whether with coconut, algae or jatropha oils have used blends with Jet-A1 kerosene.

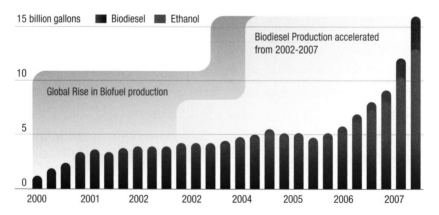

Figure 9.12. Global biofuel production is rising rapidly with ethanol commanding the greatest share of the market. Biodiesel production has accelerated in the last 5 years but is limited by availability of arable land and relatively low yields on oil crops. (Source: IEA)

Country	Feedstocks		2007 Production Forecast (*million gals*)		Blending Targets
	Ethanol	Biodiesel	Ethanol	Biodiesel	
Brazil	Sugarcane, soybeans, palm oil	Castor Seed	4,966.5	64.1	25 percent blending ratio of ethanol with gasoline (E25) in 2007; 2 percent blend of biodiesel with diesel (B2) in early 2008, 5 percent by 2013.
Canada	Corn, wheat, straw	Animal fat, vegetable oils	264.2	25.4	5 percent ethanol content in gasoline by 2010; 2 percent biodiesel in diesel by 2012.
China	Corn, wheat, cassava, sweet sorghum	Used and imported vegetable oils, jatropha	422.7	29.9	Five provinces use 10 percent ethanol blend with gasoline; five more provinces targeted for expanded use.
EU	Wheat, other grains, sugar beets, wine, alcohol	Rapeseed, sunflower, soybeans	608.4	1,731.9	5.75 percent biofuel share of transportation fuel by 2010, 10 percent by 2020.
India	Molasses, sugar cane	Jatropha, imported palm oil	105.7	12.0	10 percent blending of ethanol in gasoline by late 2008, 5 perfect biodiesel blend by 2012.
Indonesia	Sugarcane, cassava	Palm oil, jatropha	--	107.7	10 percent biofuel by 2010.
Malaysia	None	Palm oil	--	86.8	5 percent biodiesel blend used in public vehicles; government plans to mandate B5 in diesel-consuming vehicles and in industry in the near future.
Thailand	Molasses, cassava, sugarcane	Palm oil, used vegetable oil	79.3	68.8	Plans call for E10 consumption to double by 2011 through use of price incentives; palm oil production will be increased to replace 10 percent of total diesel demand by 2012.
United States	Primarily corn	Soybeans, other oilseeds, animal fats, recycled fats and oil	6,498.7	444.5	Use of 7.5 billion gallons of biofuels by 2012; proposals to raise renewable fuel standard to 36 billion gallons (mostly from corn and cellulose) by 2022.

--*negligible*

Table 9.2. Biofuel Blending Targets, Selected Countrie (Sources: FO Licht; USDA)

Referring to Figure 9.12, clearly, the trend is one of increasing biofuel production over the past few years. It is reasonable to assume this accelerating trend will continue for some time. Indeed, if countries are to have any hope of meeting the ambitious goals they have set themselves for replacement of traditional hydrocarbon fuels and reduction of greenhouse gas emissions, they will have to do much, much more. Table 9.2 shows the target levels of biofuel blending with conventional fuels for a few selected countries for 2007.

Unfortunately, however, our ability to grow sufficient crops to produce biofuels is limited. The use of arable land to grow crops for fuel is difficult to justify when large swathes of the world's population still go hungry. Even for those who can afford to pay, the diversion of food crops, like corn, into ethanol production drives food prices up and environmental groups are concerned about high utilization of water, fertilizers, pesticides and fossil fuels in the growing, harvesting and processing of biofuel products. Indeed, so-called "life cycle" calculations that consider the total energy consumed to produce different types of biofuel do not present a very favorable picture of ethanol derived from crops like corn. They suggest that energy expended (usually from fossil fuel) in the production of ethanol from corn is almost the same as that produced from the ethanol itself, with some declaring a positive balance, others a negative balance, depending on a variety of factors (e.g., distance between field and ethanol plant, yield per hectare, etc). It should be noted here that the life cycle calculations for ethanol derived from sugar cane, as in Brazil, are significantly different. Brazilian ethanol shows a very positive energy balance yielding around 8.2 times the energy required to produce it. Further developments, including the introduction of improved strains of sugar cane with up to 80% higher sugar content, should be able to make additional efficiency gains.

Compared to ethanol, biodiesel has a much higher energy density but it requires a lot of land to produce a ton of vegetable oil. The oils produced are themselves already widely used in food and industrial products so once again their use as fuels can cause spiraling prices. Often, non-edible vegetable oils, like jatropha, are promoted due to the hardy nature of the plant and the fact that it grows well, even in poor soils that would otherwise be unusable. However, recent studies suggest that, in such circumstances, the plant will yield much lower harvests than when cultivated in good soils. Thus, rather than the oft-quoted figure of 2.7 tons/hectare, yields are more likely to be closer to 1.7 tons per hectare, which translates to just over 100 gallons of biodiesel per acre per year.

Another possible source of oils for biodiesel production, and one that doesn't compete with agricultural land, are marine algae.[33] These green photosynthesizing algal communities, a kind of glorified pond scum, can be grown in seawater ponds or in suspended plastic bioreactors to provide maximum access to sunlight and, given the right conditions, can multiply rapidly. Feeding them and keeping them warm with carbon dioxide and waste heat from power plants or similar industrial sources, for example, accelerates their growth while simultaneously helping to reduce emissions. They produce an oil that can be extracted, refined and either used alone as a fuel or blended with more conventional fuels.

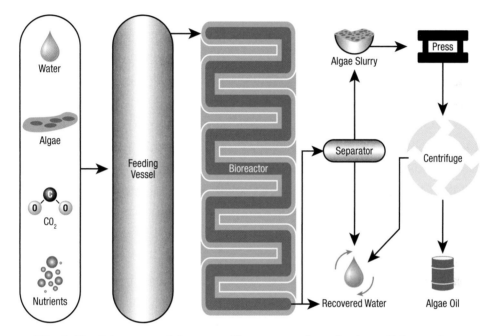

Fig. 9.13. Diagram of the overall process of producing oil from algae

Under favorable conditions, algae can produce high yields of oil – some companies estimate yields of 85 to 170 tons per hectare per year (equivalent to 10,000 to 20,000 gals per acre per year) but some experts familiar with the specific challenges posed by commercial scale-ups of this type of technology, suggest this is overambitious. They claim a more realistic figure would be a yield of around 36 - 42 tons per hectare per year. However, this is still very much better than the quantities of oil available from the more familiar land-based crops.

Growing crops, whether for fermentation or oil extraction, is not the only method that can be employed to produce liquid fuels from renewable sources. Animal fats can produce usable biodiesel from the same transesterification process used with vegetable oils and, indeed, such an approach is used, at least to some extent, in countries with large meat processing industries. In more recent developments, Tyson Foods and ConocoPhillips have entered into a partnership together to produce biodiesel from some of the 2.3 billion pounds of waste animal fat generated by Tyson's rendering plants each year. In this case, however, the waste fat is processed in a conventional refinery and could, theoretically, yield 300 million gallons of biodiesel.

A rather more versatile technology for producing fuels is thermal depolymerization, which can convert any organic matter, animal or vegetable, into oil. The technique involves heating the feedstock to around 800 °C under pressure and in the absence of air. Under such conditions, organic materials (cellulose, proteins, fats, sugars, etc and even waste plastics) decompose to yield an oily product that can be refined in a conventional refinery or can be used, as is, to fire industrial boilers. One company had planned to use this technique at its plant in Missouri to transform 78,000 tons of waste per year mainly from a nearby turkey processing factory into 9 million gallons of fuel.[34] However, the economics of the process require oil prices above about $80 per barrel to justify it and while the prototype plant has shown proof-of-concept for the technology, lower oil prices are likely to undermine its commercial viability, at least in the short term.

1. Prepare the organic and inorganic feed with water.

2. Heat the slurry under pressure to reaction temperature.

3. Depolymerization reaction: Separating the organic from the inorganic.

4. Hydrolysis reaction: Water splitting complex molecules into smaller units.

5. Inorganic material to solid storage

6. Separation of gases, renewable diesel, water and remaining solids.

7. Renewable Diesel storage.

Figure 9.14. Thermal Conversion Process

While not truly "alternative" fuels, from the perspective that they are still derived from traditional and finite hydrocarbon sources rather than renewable sources, it is worth mentioning some other processes that are already in use for the production of liquid transportation fuels. These are the far more attractive GTL and CTL or, respectively Gas-to-Liquids and Coal-to-Liquids and, as their names suggest, these processes convert natural gas or coal into liquid fuels. Both techniques are based on methods, for the most part, developed in the early twentieth century by German chemists. In the case of GTL, natural gas (mainly methane) is converted to a mixture of carbon monoxide and hydrogen (syngas) by a process called steam reforming. This mixture is then reacted under temperature and pressure in the Fischer-Tropsch process, using selective catalysts based on iron or cobalt, to produce liquid paraffin hydrocarbons. These may be used directly, or further treated by hydrocracking, to produce gasoline, non-aromatic diesel and naphtha.

Coal-to Liquids technology can make use of the same Fischer-Tropsch synthesis if the coal is first gasified by treating it with superheated steam. Alternatively, the coal may be finely ground and hydrogenated by using hydrogen gas or by dissolving it in a solvent, mixed with a molybdenum catalyst under temperature and pressure. South Africa has been producing liquid fuels from coal since 1955 and has a production capacity of 160,000 bpd, producing around 30% of the country's gasoline and diesel requirements, in this way. While there are concerns about the sustainability of such schemes and about their environmental credentials, the commercial and technical success of these methods prove beyond doubt that they are practical. As pointed out in Chapter 7, coal is the most abundant hydrocarbon on the planet and large deposits exist in many countries regarded as friendly or politically stable.

BioMass-to-Liquids (BMTL) is one alternative approach that has garnered increasing attention. Using essentially the same approach as GTL and CTL, biomass and other organic waste is converted into syngas, which is then used in the Fischer-Tropsch process to produce liquid fuels. The advantage of BMTL over other synthetic schemes is that it uses renewable source materials rather than hydrocarbons like coal or gas. Apart from the standard Fischer-Tropsch approach, alternative BMTL methods include thermal depolymerization (described above) and flash pyrolysis. In the US alone, the Departments of Energy and Agriculture have estimated 1.3 billion tons of biomass waste could be collected per year without affecting food production. This could be used to make an estimated 100 billion gallons of fuel using BMTL and other conventional fermentation technologies.

Of course, transportation fuels, for surface vehicles in particular, need not necessarily be liquid; certain gases may be used, too. Hydrocarbon gases like propane and butane have been used to power fleet vehicles and taxis for many years in some countries and have become more widely available for normal cars in recent years. Although gaseous under normal temperature and pressure conditions, these fuels are stored as liquids at relatively low pressure (~300 psi or 22 bar). While either gas can be used alone, more commonly mixtures of both are used and these mixtures are generally referred to as Liquefied Petroleum Gas or LPG. It is estimated that as many as 13 million vehicles worldwide use LPG, consuming about 7 billion gallons per year.

Even more interesting, and becoming ever more practical based on a new appreciation of the true magnitude of global reserves, is the possibility of using the principal component of natural gas, methane, as a fuel for vehicles. This is already being done in many countries but a massive expansion in the use of this fuel represents a truly fabulous opportunity to radically change the world's energy landscape. Given that 70 percent of crude oil is used to produce transportation fuel (primarily gasoline and diesel), a switch to natural gas for transportation would have an enormous impact. In the US, for example, using such a strategy would reduce crude oil consumption and, more importantly, potentially provide energy independence by replacing up to 15 million BOPD. With current US consumption of 21 million BOPD, the difference of a mere 6 million BOPD could be largely met by US domestic crude oil production. At the same time, based on 2010 oil prices, it would cut import costs by some $450 billion a year, providing a much needed boost to the country's balance of payments. Some of this money would still be spent paying for the required natural gas fuel but that spend would be largely internal to the US economy. The surplus could be spent on government subsidies to encourage the switch to natural gas vehicles and on upgrading the infrastructure to enable such vehicles to be refueled anywhere in the country.

When used as a transportation fuel, natural gas is usually compressed and stored in high pressure cylinders at pressures ranging from 2500 – 3500 psi (170 to 240 bar) and referred to simply as Compressed Natural Gas or CNG. The fuel can be used in any conventional internal combustion engine after suitable conversion and some commercial models are available that can be run on both CNG and gasoline. Recently, progress has been made on efforts to store the natural gas in cylinders filled with high specific area adsorbents like activated carbon and Metal Organic frameworks (MOFs). Advantages of such an approach include the use of much lower pressure cylinders, saving weight and improving safety. Implicit, too, in this approach, is the even more attractive possibility of refueling with gas directly from existing low pressure commercial pipeline infrastructure, without the need to provide additional compression. This would greatly simplify nationwide implementation. While not yet perfected, this technology has already demonstrated storage capacities that approach the targets specified by government scientists (180 vol/vol at 35 bar and near ambient temperature).

The ultimate aim of many is to eliminate carbon from transport fuels altogether. One way to do that would be to switch to hydrogen. The latter, when burned in a combustion chamber, or better yet combined with oxygen in a fuel cell, produces only water as its reaction product and hence eliminates the perceived threat posed to the world's climate by continued emissions of carbon dioxide (CO_2). Serious concerns and challenges remain, however, about the technical feasibility of using hydrogen in this way since it is extremely flammable, potentially explosive across a wide range of concentrations in air, and must be stored in high pressure cylinders or at extremely low temperatures like LNG. Efforts are ongoing to improve storage technology for hydrogen using adsorbents and this would certainly make it more attractive as a fuel. Even more problematic is the low energy density of hydrogen. A gallon of liquid hydrogen contains only 29% of the energy contained in a gallon of gasoline (2.6 kWh per liter vs. 9.0 kWh per liter) so fuel tanks would need to be proportionately larger in relation to the size of the vehicle (and would need to be cooled to -253 C, or -423 F, too!).

Hydrogen for use as a fuel is made today by "reforming" of natural gas, an intrinsically inefficient process. Indeed, making hydrogen in this way uses more energy than would have been obtained by simply burning the gas, in the first place. Hydrogen can also be made electrolytically by passing an electric current through water and splitting it into its constitutive elements (i.e. hydrogen and oxygen). However, again, the process consumes more energy than that provided by the hydrogen produced. It may be viable to use this approach assuming that the electricity in question comes from solar or other non-carbon emitting sources but, even then, it makes little sense to go to the trouble of producing, distributing and selling hydrogen.[35] It would be simpler, cheaper and more effective to use the electricity directly to charge an electric vehicle rather than to make hydrogen. And this last possibility, once deemed too challenging to be a serious contender, is ultimately what may prove to be the best alternative fuel for mass-market road vehicles, after all. Advances in a variety of areas make this all the more likely.

Electric vehicles are not new. In fact, they pre-date vehicles powered by internal combustion engines (ICEs) by at least half a century and were popular, in both Europe and the US, until around 1920. Their demise can be traced to the discovery of vast reserves of oil in Texas, the arrival of cheap mass produced ICE vehicles by Henry Ford and the increasing network of roads for long distance travel in the US. With the price and availability of gasoline making it easily accessible to consumers, electric vehicles could compete neither on range nor on cost with the mass-produced ICE vehicles. The cost of the latter continued to fall and by the late 1930s, the electric automobile industry had, to all intents and purposes, disappeared.

Figures 9.15 and 9.16. Early electric vehicles

Excluding hobbyists and captive urban utility fleet vehicles, electric cars were essentially forgotten for most of the remainder of the twentieth century. On occasion, they re-appeared as concept vehicles at motor trade events but they were never considered worthy competitors to the endless range of conventional vehicles, priced to suit every pocket, powered by the ubiquitous internal combustion engine.

Figure 9.17. The General Motors EV1 Electric Roadster was unveiled at the Los Angeles Motor Show in 1990

However, variants of the electric vehicle began to appear again, probably in response to impending legislation that sought to force manufacturers to develop them, in the early 1990s. The EV1, a concept electric roadster was unveiled at the Los Angeles Motor Show in 1990. But, it was not until the arrival of the commercially available Toyota Prius hybrid sedan in 1997 that electric vehicles really started to be taken seriously. Hybrid vehicles, which combine a conventional gasoline or diesel engine with an electric motor and battery pack, have already demonstrated some of the possibilities of, improving fuel efficiency significantly. The Toyota Prius, to many the signature hybrid vehicle and certainly the most successful commercially, demonstrates efficiency improvements of some 30 - 40% over conventional vehicles, achieving 50 miles per gallon (mpg) compared to 30 to 35 mpg for many standard vehicles. The Toyota Prius is a "parallel-hybrid" vehicle in which the vehicle alternates between being powered either

Figure 9.18. From top to bottom, the Toyota Prius, Honda Insight and Tesla Roadster. The latter is an all-electric vehicle powered by Lithium-ion batteries, the others are hybrid vehicles, featuring battery pack, electric motor and internal combustion engine.

by the electric motor or by the internal combustion engine. An alternative type of hybrid vehicle is the "series-hybrid" in which there is no connection between the internal combustion engine and the drive train. Instead, the ICE is used strictly as a generator to provide electrical power to the motor and to charge the battery, as required. While these early hybrid vehicles are not designed to be plugged-in and recharged (charging is normally accomplished at higher speeds when the internal combustion engine is running and by harvesting energy from "regenerative" braking), enthusiasts have already modified some hybrids so they can be charged from the mains supply. In so doing, they are able to access fuel, in the form of electricity, that has a retail cost currently equivalent to 25 cents per liter of gasoline (i.e. under $1 per US gallon equivalent). Of course, in order to achieve this, these enthusiasts have had to install additional batteries to extend range and improve performance but they have shown, at least, the feasibility of the approach. Close to 2 million hybrid vehicles were in service by early 2009, and several new models are pending, most with the optional facility, this time, to be recharged from the mains (so-called Plug-In Hybrids).

As noted above, until fairly recently, electric vehicles were considered impractical. Attempts over the years to produce such vehicles for commercial use met initially with skepticism or even ridicule. In general, these initial efforts relied on traditional battery technologies like lead-acid cells, of the sort used in ICE ignition systems. Such batteries are far from ideal due to their weight, long charge times and relatively slow release of stored energy. While they can provide a high current surge to start a conventional engine, they cannot sustain such loads and are susceptible to deep discharge problems, whereby they suffer permanent impairment if discharged beyond a certain point. Many motorists are only too familiar with this effect, having tried in vain to start a vehicle on a cold, frosty morning. Attempts to start the cold engine drain the battery quickly and it may never recover its capacity to hold charge again. Nickel/Cadmium batteries were also used in early electric vehicles but concerns about pollution and long-term performance limited their application.

In the 1990s, the newer, more efficient nickel metal hydride batteries became widely available. These were relatively cheap, used less toxic materials and, most importantly, could be recharged almost indefinitely, unlike their predecessors. Thus, the batteries could potentially outlive the normal lifespan of the chassis eliminating the need to replace them during the vehicle's expected life. Finally, Lithium ion batteries have entered the race, featuring in such vehicles as the Tesla Roadster. Li-ion batteries are light and can hold high charge densities, making them ideal for use in electric or hybrid electric vehicles. Again, they can be discharged and recharged many times, without significant deterioration, and have estimated life spans of around 300,000 miles (nearly 500,000 kilometers). Early versions of these batteries were prone to overheat but newer designs using lithium iron phosphate are more stable and have shown promise in addressing the last major obstacle for the widespread use of electric vehicles – the problem of re-fueling.[36]

The convenience of filling-up a conventional gasoline vehicle in a matter of minutes is something that we take for granted today. So, too, is the distance we can then travel on a full tank of fuel - typically 200 to 350 miles (~320 to 560 kilometers). Similar practices can be employed with fuels like biodiesel or butanol or DMF, in appropriate vehicles. Pure ethanol, on the other hand, presents problems due to its lower energy density necessitating more frequent stops to refill the tank. The problem is even worse for hydrogen since it has 70% less energy per unit volume than gasoline and, with its wide flammability limits and its volatility, it would require far more stringent safety measures than any of the other fuels mentioned. All of these fuels, including gasoline, have to be delivered by road tanker or, perhaps by pipeline in certain cases, requiring investment in equipment, facilities and infrastructure, thereby increasing cost.

In the case of electric vehicles, however, the "fuel" can be delivered via the electrical distribution grid. While not perfectly efficient, long range electrical transmission losses are relatively low and accurate systems are already in place to measure consumption of electric power (or fuel). This has always been feasible for vehicles that are topped-up or recharged with power from the mains. However, until recently, this process of recharging was time-consuming, requiring several hours or, typically, being done overnight. That may be acceptable for vehicles used in urban settings where the daily distance traveled is short but it would be difficult and inconvenient, to say the least, for longer-distance travel between cities, for example. Having to stop for several hours to refill with fuel would be unacceptable. To date, the best way around this problem for electric vehicles was the hybrid vehicle. As noted above, however, the latest developments in battery technology may have solved this. Researchers have reported a novel method to prepare batteries containing finely divided lithium iron phosphate. The resultant high surface area and electrical properties of the composite allow it to discharge and recharge very quickly. Whether it lives up to expectations remains to be seen but reports that it may be possible to recharge an electric vehicle battery pack based on this material in a minute or two, without risk of overheating or explosion, are certainly encouraging.

Such a development, if realized, would be a major step towards making a ground transportation system based on electrical power feasible. Clearly, there are still many challenges to the implementation of such a scheme.

Improvements in storage technology may also be required to make it feasible for "filling stations" to dispense power at the rate needed to recharge a vehicle in a matter of minutes. Much progress has been made in recent years in developing ultra-capacitors, a type of electrical storage device that can charge and discharge much faster than batteries. These devices are already in use in some electric vehicles to provide extra power for acceleration that batteries cannot deliver today, due to their slower discharge rates. It is possible that large banks of these ultra-capacitors could act as reservoirs for electrical energy at "filling" stations. They could charge continuously from the conventional grid yet would be able to discharge quickly to refill a vehicle battery pack in a very short time, perhaps. Alternatively, such "filling stations" may need special connections to the grid, which in all likelihood will itself need to be upgraded to handle the extra demand imposed by electric vehicles, in the future. Of course, many vehicles could still be charged at a more leisurely pace through the night, while plugged-in at home, using off-peak electricity, or during the day at supermarket car parks or while parked at work. Thus, the need for dedicated "filling stations" is likely to be less than what is currently required to dispense traditional liquid fuels.

We began this chapter by dismissing the risk that the world would run out of conventional hydrocarbon fuels anytime in the near future and it is fitting that we should return to that same point to close it. We have used hydrocarbons, as solid, liquid or gas, to fuel our industries, our homes and our vehicles for several centuries now. We did so because they were widely available, relatively energy-dense and cheap. We will continue to use hydrocarbons for some of those same uses and for those very same reasons for the foreseeable future and we might even see a significant increase in the use of gas, for example. However, now, we also have the very real option to use a variety of alternative fuels, derived from a diverse inventory of source materials. Not all of these may prove to be ultimately successful as fuels and a few will probably fall by the wayside. The reasons for failure of some of these materials will be technical, or practical, or related to safety, perhaps. For others, it will be because they are not commercially viable or because they compete with other equally important resources, like water or arable land, for example. As this century progresses, however, we are likely to see more and more emphasis being placed on those that show real promise as replacement for petroleum fuels. The net effect of these alternative fuels, however, will not be to end the age of petroleum but, rather, to extend it.

CHAPTER 10

Pollution and the Environment

Chapter 10
Pollution and the Environment

Unquestionably, the oil and petrochemical industries have been key drivers in the development of the complex, modern world we live in. As we have discussed in several places in this book, hydrocarbons have helped power cities, transport goods and provide the raw materials for the manufacture of countless other useful materials including pharmaceuticals, fertilizers, dyes, plastics and consumer products. As noted in Chapter 3, many of the advances made in general health, and in improved living standards around the world, can be traced to the wealth generated by widespread access to affordable energy.

The population of the world stood at approximately 6.8 billion people in 2010.[37] This falls far below the population growth projections made by various prophets of doom since as early as the eighteenth century but it is, nonetheless, a formidable number of people. All of these people need food, water and other resources, including energy supplies. They also produce substantial quantities of waste, some organic and easily recyclable, but a variable and increasing portion of this waste must be treated on an industrial scale to render it suitable for discharge to the environment. In the so-called developed world, the industrial revolution provided the impetus for the rapid implementation of labor-saving technologies and the widespread introduction of machines. The economies of scale and efficiencies made possible by these machines were key elements in the dramatic improvements in health and living standards that are taken for granted today in many parts of the world.

For the fortunate minority in the developed world, life is no longer a hardscrabble struggle to survive, so it is not surprising that there should be increasing recognition of the potentially detrimental effects of some of mankind's activities on the environment.

In 1962, Rachel Carson published *Silent Spring*, a book that highlighted the potentially toxic effects of man-made pollutants on the biosphere.[38] The "silent spring" she referred to was one bereft of songbirds, which Carson claimed were increasingly being harmed by pesticide residues, primarily DDT, a powerful synthetic insecticide that had been discovered in the 1930s. Since DDT had no apparent toxic effects on higher life forms, including humans, it was widely and perhaps indiscriminately used. However, it became clear that DDT and similar molecules could persist for a long time in the environment, and owing to their solubility in body fat, DDT levels could gradually build up in the bodies of birds and mammals in a process called bioaccumulation. Carson's suggestion that this could cause cancer and birth defects in humans as well as animals, ultimately, led to a ban on the use of DDT. Today, many believe that banning DDT has had major deleterious effects on the world and that millions of people may have perished unnecessarily because of diseases borne by mosquitoes and other insect vectors, or because of famines caused by agricultural pests.[39] However, those who have challenged Carson's and others' claims regarding DDT toxicity rather miss the point of environmental activism.

Carson's work was important, but not necessarily for its accuracy. Instead, its strong emotional impact, irrespective of scientific validity, served as a clarion call for individuals and groups with similar social and political inclinations and provided a rallying point to unify many disparate interests. Ultimately, this gave rise to the environmental movement, typified today by organizations such as Greenpeace, Friends of the Earth, the Sierra Club and many others. Although some of these groups and individuals within these groups have acted responsibly and taken responsible positions on important issues, others have not. Worse yet, a few have acted in profoundly non-progressive ways and, in certain cases, in ways that would harm the common man and the welfare of many people in the world.

The activities of environmental groups should be more globally beneficial than harmful, but despite the considerable societal good that hydrocarbons generate, these groups have had a rather uneasy and equivocal relationship with the energy industry. As the ultimate provider of the raw materials used in the manufacture of many organic chemicals, including pesticides like DDT, the hydrocarbon industry has been demonized and targeted systematically by environmental groups for many years. Also, because these groups consider burning hydrocarbons as fuel to be largely responsible for increases in atmospheric carbon dioxide, they view the industry as the principal culprit behind supposed Anthropogenic Global Warming (AGW). Finally, of course, they accept zero tolerance for environmental effects related to the search for, transportation of, and refining of oil and gas.

Many people who are not environmental activists worry about the risks posed by exploration for and exploitation of hydrocarbon deposits because, in the public mind, crude oil represents one of the most acute and visible threats to the environment. As we have seen in earlier chapters, its commercial exploitation involves tapping reservoirs capable of producing at the prodigious rates needed to supply worldwide consumers. In 2010, world oil production averaged some 85 million barrels of oil per day, and reliable sources estimate that this will rise to around 120 million barrels of oil per day by 2050. As a major world commodity whose source is often far-removed from the end-user, oil is transported around the globe in vast quantities. Thus, any serious accident tends to involve discharge of substantial volumes of oil.

The occasional but highly-publicized spills can have distressing effects on wildlife. Most everyone has seen the aftermath of such events on the news – oil-soaked marine mammals, pathetic sea birds, dead fish and a blackened, uninhabitable coastline. The *Exxon Valdez* tanker accident in Alaska in March 1989 is one of those incidents along with the deliberate release of oil into the Arabian Gulf and the systematic destruction of wellhead control equipment in Kuwait by retreating Iraqi forces at the end of the first Gulf war in 1991. But probably the most widely known incident today, and the one that has created the greatest public and political furor in recent times, is the blowout of BP's deepwater Macondo exploration well in the Gulf of Mexico in April 2010.

The accident occurred while the rig *Deepwater Horizon* was drilling an exploration well in a water depth of over 5,000 feet. Drilling wells in such deep water is a relatively new development, although the first proof-of-concept wells at such depths were drilled by the French company Elf Aquitaine in 1983. For reasons that are unclear as this book goes to press, the BP well experienced a serious "kick" – a high-pressure gas eruption that made its way to the surface, causing a fiery explosion and resulting in the tragic deaths of 11 crew members and the sinking of the rig. Multiple independent safety systems that should have closed the well at the seabed failed to operate as designed, leaving the well unsecured and leaking oil into the Gulf of Mexico. The flow rate from the well was initially estimated to be up to about 5,000 barrels per day, but the figure was subsequently revised up to between 12,500 and 20,000 barrels per day and some estimates put this as high as 70,000 barrels per day.

A well leaking at such rates is, obviously, a major incident and all the more so since the rig was only 40 miles (64 kilometers) from the US coast. The contamination of major fisheries, wildlife conservation areas and tourist beaches in several states was, therefore, practically certain. With the Blowout Preventer (BOP) -- the series of hydraulic valves that should have closed the well -- apparently non-functional and wreckage of the sunken rig and marine riser strewn around and on top of the seabed wellsite, closing the well proved extremely difficult. Human divers cannot operate at such depths and ROVs (Remotely Operated Vehicles or small submarine robots) were unable to activate the BOP. As oil continued to spew from the well, various methods were tried to capture or re-direct the leaking oil or to siphon it from the flow stream and prevent it from entering the sea. These included lowering a giant, upended steel funnel over the well and inserting a small siphon pipe into the damaged well and cutting and partially capping the leaking riser pipe. While the latter met with some degree of success, oil continued to pour into the sea, day after day. The American public, sensitized by many years of increasingly strident environmental activism against oilfield activities, and against offshore drilling in particular, went into shock.

Figure 10.1. Transocean's rig Deepwater Horizon was destroyed and 11 crew members killed when BP's Macondo exploration well blew out in the Gulf of Mexico. Inability to close the well on the seabed in over 5,000 feet of water resulted in an oil leak estimated to be between 50,000 to 70,000 barrels per day.

The incident and the sensational 24-hour media coverage set the scene for an environmental spectacle that bordered on circus: ubiquitous finger-pointing by BP, Transocean (owners of the sunk rig), Halliburton (the service company that cemented the well), Cameron (the BOP supplier) and the Minerals

Management Service (MMS), the industry regulatory body that grants drilling permits; an industry that seemed to have cut corners on safety and had no idea how to control the leak; and a US administration that withdrew plans to extend offshore drilling into previously off-limits areas of the US coastline. People demanded action, and the government threatened to take control of the situation before reluctantly admitting it had no expertise and had to rely on BP's (and other industry players') technical acumen.

Figure 10.2. The Pemex Ixtoc-1 blowout in 1979 released a total of 3.3 million barrels of crude oil into the Gulf of Mexico over a 10 month period.

For the American public, the *Deepwater Horizon* incident was certainly a wake-up call. They watched, understandably horrified, as subsea cameras showed oil pouring from the stricken well and aerial and satellite images showed the ever expanding oil slick in the Gulf of Mexico. For many, this was an unprecedented and almost apocalyptic event. The blogosphere erupted with calls to punish BP for their perceived negligence, to forbid the company from further work in the US, and to seize their US assets. Few recognized that in 2010, BP was the biggest oil producer in the Gulf of Mexico and for many years previous was the biggest oil producer in the entire United States. Leakage from the well was, inevitably, benchmarked against the oil spill from the *Exxon Valdez* tanker, mentioned above, and found to be much more severe than that prior benchmark.

The *Exxon Valdez* spilled an estimated 250,000 barrels of crude oil into the waters of Prince William Sound in 1989. When a specially engineered containment cap was successfully installed on the BP Gulf of Mexico blowout on July 15, 2010, effectively shutting in the leaking well, estimates of the total spill volume since it began on April 20, 2010 ranged from about 1.8 million barrels to over 4 million barrels. Depending on which estimate is regarded as valid, this made the BP spill potentially larger than the previous record blowout in the Gulf of Mexico, the *Ixtoc-1* well, operated by Mexican state oil company, Petroleos Mexicanos (Pemex), which flowed some 3.3 million barrels of crude oil into the Gulf in 1979.[40] Oil from that blowout, which was finally brought under control after 10 months, contaminated the beaches of South Texas.[41]

Both of these blowouts were relatively insignificant, however, when compared with the oil that was released deliberately by Iraqi forces retreating from Kuwait in 1991. Those forces sabotaged over 600 oil wells with explosives.[42]   While impossible to assess accurately, it is estimated that the total amount of oil released amounted to around 1 billion barrels, of which perhaps 60 million barrels remained uncombusted, and about 11 million barrels discharged directly into the turquoise-blue waters of the Arabian (Persian) Gulf.[43]

Figure 10.3. Retreating Iraqi troops sabotaged more than 600 Kuwaiti wells in the 1991 Gulf War, destroying wellheads with explosives and causing the release of an estimated 900 million barrels of crude oil.

Several other large accidental spills have caught the world's attention, although they were not, of course, in the Gulf of Mexico and hence went largely unnoticed or have been long forgotten by the American public. The *Torrey Canyon* and the *Amoco Cadiz* were two tankers that sank in or around the English Channel (actually off the Brittany coast) in 1968 and 1978, respectively. The *Torrey Canyon* was carrying 120,000 tons (860,000 barrels) of crude oil and the *Amoco Cadiz*, 219,000 tons (plus 4000 tons of heavy fuel oil) or over 1.6 million barrels. In the latter case, the entire cargo was released into the sea within a few days, contaminating over 200 miles of the French coastline.[44]

Figure 10.4. The Amoco Cadiz tanker spill released 1.6 million barrels of crude oil, contaminating 200 miles (320 kilometers) of the French coast.

The aftermath of such events, as portrayed by the media, is always particularly harrowing, and the public have been left with a very clear impression of the damage a crude oil spill can do. Sadly, what the media rarely note is that oil is a completely natural product, produced by the Earth's own natural processes. It is also quite naturally and normally degraded by a variety of organisms and the environment itself. Hydrocarbons routinely seep to surface in a variety of locales, including large areas of the Gulf of Mexico, through normal geological processes. Many of the earliest oilfields were actually first identified by such phenomena. Under certain circumstances, these hydrocarbons provide a rich nutrient source for various creatures.[45] It is only when large volumes of oil are discharged in inappropriate places that it poses a pollution problem because it swamps the environment's ability to degrade it in the short term. Because oil is, by nature, hydrophobic (water-repelling), the organisms that live in water have difficulty accessing this rich source of energy, initially. However, when it is properly dispersed, it can be degraded relatively quickly. For example, in January 1993, the *Braer*, a Liberian-registered tanker carrying Norwegian crude oil to a refinery in Quebec, was caught in hurricane force conditions off the northern coast of Scotland. She ran aground, the hull was breached and the majority of her 85,000-ton cargo escaped. Despite valiant efforts by all concerned, little could be done for many days due to the ferocity of the storm. Three weeks later, there was remarkably little evidence that the *Braer* had sunk. The natural power of the wind and waves had done an impressive job of dispersing the crude; for all intents and purposes, it had vanished. The official investigation report concluded[46] :

"The official death tolls included 1542 seabirds, several thousand pounds of commercially farmed salmon, 10 seals and 4 otters. Two of the otters were run over by a camera crew covering the spill, however; and the other two probably died of old age"

In the case of the *Exxon Valdez*, numerous commentators have trotted out what sound like rather grim statistics. The toll of animal deaths associated with the spill included an estimated 250,000 seabirds, 300 harbor seals, 2800 sea otters, 250 bald eagles and, possibly, 22 killer whales. While these numbers sound dreadful, they suggest more than they reveal. As a comparison, more than 250,000 birds die in the US every day in collisions with plate glass windows.[47] Similar numbers of birds are killed every two days in the UK by domestic cats. And, while the comparison is not really relevant, over 20 million chickens are deliberately slaughtered for food each day in the United States with few consumers being aware of the fact[48], and even less caring about it.

We are not for a moment suggesting that the world should be complacent about the effects of oil spills. In almost any incident involving uncontrolled release of hydrocarbon, errors in judgment or poor decisions that contributed to the event can usually be identified, after thoughtful investigation. The effects of such errors, had they been foreseen, could usually have been mitigated by better engineering or better legislative controls. However, it is important to balance hysteria and exaggerated commentary about such events with facts.

For example, although the Deepwater Horizon incident is surely the biggest offshore oil spill in the history of the US, it falls far short of much bigger spills on American soil. In the early days of oil exploration, blowouts (or "gushers" as they were then known) were common. The Spindletop well in Beaumont, Texas probably flowed at rates of 100,000 barrels per day and released at least a million barrels of crude oil into the environment. The most famous of these wild wells was the 1910 Lakeview gusher in California, which flowed uncontrollably at rates of 18,000 to 100,000 barrels per day for almost 18 months, spewing out over 9 million barrels of crude oil, most of which soaked into the soil.[49]

Figure 10.5. Natural seeps in many locations leak oil and gas into the biosphere. It is estimated that around 4.3 million barrels of oil leaks naturally into the ocean annually from such sources and such seeps may leak for thousands of years. The above natural leaks are in the Gulf of Mexico.

In addition, oil and gas are constantly entering the environment and have been doing so long before mankind thought to start drilling for them. According to the National Academies of Science, a total of 1.3 million tons of oil enters the oceans each year. Of this, an estimated 600,000 tons (~4.3 million barrels) or about 45 percent of the total originates from natural seeps. These seeps are natural conduits – fractures, faults or high-permeability streaks in the rock strata that lie above oil and gas reservoirs – and they allow hydrocarbons to escape to surface. In US waters alone, 160,000 tons (more than one million barrels) of crude oil leaks into the sea from such seeps, primarily in the Gulf of Mexico and off the California coast.[50] In one of the best known areas, Coal Oil Point, off Santa Barbara, 100 to 150 barrels of crude oil and approximately 3.5 million scf of natural gas seep into the sea each day. This equates to about 55,000 barrels of oil and 1.3 billion cubic feet of gas per year – or an *Exxon Valdez* every 5 years. Such seeps have likely persisted for thousands of years, and measurements of residual hydrocarbons in marine sediments around the Coal Oil Point seep suggest they may contain as much as 80 "*Exxon Valdez* equivalents" of oil.[51]

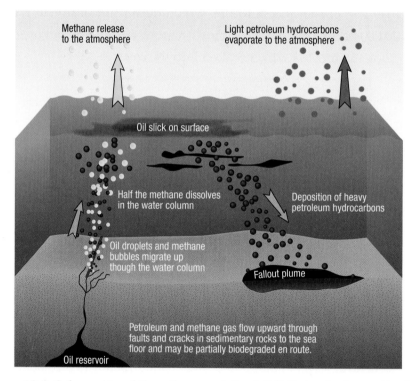

Figure 10.6. Schematic of the oil seep at Coal Oil Point in Santa Barbara. Approximately 55,000 barrels of oil and 1.3 billion cubic feet per year of gas leak from the seabed each year. (Source: Woods Hole Oceanographic Institution)

Researchers have also noted that oil exploration and production probably reduces the leakage of hydrocarbon from such seeps into the environment by reducing reservoir pressures. Also, the NAS estimates that, under normal circumstances, upstream oil and gas activity (drilling, production) is directly responsible for an average of only 38,000 tons of oil input to the oceans, worldwide, per year. This equates to less than 3% of the total oil input to the sea.[52]

Knee-Jerk Reaction

Predictably, as a result of the BP blowout, many groups and individuals called for the suspension of any further offshore oil drilling until an independent investigation of the BP spill could be completed. On April 30, 2010, the US administration acquiesced to such demands and reversed plans to expand offshore drilling, pending an investigation into the causes of the blowout. It also announced a 6 month moratorium on all deepwater drilling in the GoM, effectively cutting domestic oil and gas production, losing $700 million in taxes and royalties and putting the livelihoods of 23,000 people who had nothing whatsoever to do with the Deepwater Horizon well, in jeopardy. Such political grandstanding is akin to shutting down all air travel in the US, regardless of airline or aircraft type, in response to a single air crash involving one type of airplane.

Despite the emotive images of oily seabirds and marine mammals and the understandable sympathy for concerns of Gulf Coast residents whose livelihood may be threatened by an event like the BP spill, it is irresponsible to jump to hasty decisions about offshore drilling based on a single incident. In 2010, 30% of US domestic oil production was from offshore wells, and this percentage is bound to increase in the future. So, it is particularly worthwhile to consider the economic impact of offshore drilling, while simultaneously factoring in environmental and other costs. Fortuitously, a 2009 study by economists Robert Hahn and Peter Passell did just that. Their analysis clearly showed that the benefits of producing offshore oil greatly outweigh the costs.

The economists considered three types of benefits: production revenues, lower prices to consumers, and less volatility in oil prices. They examined these benefits in scenarios where oil was priced at $50 per barrel and $100 per barrel. At both figures, they then assessed the quantities of oil that might be economically recoverable. Logically, an increased oil price makes it more attractive to develop more marginal fields (albeit at higher cost), thereby increasing the recovery estimate. Thus, at the lower oil price they estimated 10 billion barrels of oil would be recoverable from the outer continental shelf, while at the higher oil price of $100 this was increased to 11.5 billion barrels.

Of course, there are costs associated with any production and they calculated $17 per barrel and $20 per barrel for both price scenarios. While the figure may require further scrutiny, they also used an estimate from the MMS for the cost of environmental damage resulting from the production of 10 billion barrels of oil offshore. Finally, also included were estimates of costs associated with greenhouse gas production from the use of the oil as fuel and rather more esoteric cost components like air pollution, traffic accidents and congestion. Their "number crunching" came up with some rather persuasive figures.

Assuming an oil price of $50 per barrel, the benefits of offshore oil production in the formerly off-limits areas of the US outer continental shelf produced a net economic benefit of $323 billion. At the higher oil price of $100 per barrel, the net benefits amounted to $967 billion. Hahn and Passell also noted that even if costs were to be much higher than their assumptions, the economics would still strongly favor developing these offshore resources. They considered the impact of hugely increased environmental penalty costs too, including greenhouse gas emissions that may ultimately be taxed by a cash-strapped administration up to the scandalous levels proposed by British economist Lord Stern, or which may gravitate to those same values because of financial speculation on the carbon credit market. Thus, even at an inflated carbon price of $321 per ton (in 2010, it was around $16 per ton), the net benefits of $120 billion and $725 billion, at $50 per barrel and $100 per barrel oil, respectively, would still strongly favor offshore development.

Even when other environmental impacts are incorporated into the analysis, including all the costs (socio-economic, clean-up, compensation) of a major spill, the case for offshore development is unassailable. Considering BP's misfortunes in 2010, at the time of writing, such costs for their scenario have been estimated to be as high as $20 billion. This is about ten times higher than the costs associated with the *Exxon Valdez* spill so should be considered at the higher end of the scale. Regardless of this high potential cost, and even if that figure were to be doubled, the overall impact of offshore oil production is highly beneficial.

Finally, on this point, it is important to consider one additional sobering thought. A prohibition on offshore oil and gas activity would leave the United States with a daily shortfall of around 2 million barrels of oil per day that could only be filled by increased imports. Some, undoubtedly, would come from Canada but the rest would come from West Africa, the Middle East and Venezuela by tanker – on average, an extra two supertankers per day or about 700 per year. One single, large oil tanker accident could result in a spill of similar magnitude to that produced by the BP blowout, and in a much shorter time. Accordingly, its impact would almost certainly be worse since a stricken tanker might release 1 to 2 million barrels of oil in a matter of days, rather than the weeks or months required for such volumes to be produced by a wild well. With little time available to respond to such a tanker spill, coastal damage would likely be much more severe.

While recognizing the undeniably deleterious short-term impact of large releases of hydrocarbon to the environment, the longer-term picture is much more encouraging. In the cases of those large spills studied extensively, like the *Amoco Cadiz* and *Exxon Valdez* incidents for example, scientists have found that the initial catastrophic effects are attenuated fairly rapidly. In the case of the Amoco Cadiz, it was estimated that more than 260,000 tons of marine animals (including 6000 tons of oysters) were killed by the spill. In the worst affected areas, the year immediately following the spill was characterized by a proliferation of opportunistic species, resistant to the presence of oil, which replaced the usual fauna. Over the subsequent years, these opportunistic species gradually gave way to tolerant species. By 1982-1983, four years after the spill, such tolerant species accounted for over three quarters of the marine population. Finally, as natural processes continued to degrade remaining traces of residual oil, species that were sensitive or very sensitive to hydrocarbons began to resettle and attained their normal pre-spill level by 1984-1985. In total, therefore, it took 6 to 7 years for the former balance to be regained.[53]

Figure 10.7. Numerous techniques were tried in attempts to clean up oil-fouled coastline after the Exxon Valdez spill in 1989.

The *Exxon Valdez* incident received enormous publicity and the general public's opinion of the subsequent clean-up effort was poor. The viscous crude, low ambient temperatures and creation of a thick, sticky, emulsified "mousse" that adhered to rocks made clean-up crews look ineffectual under the glare of the media. Numerous techniques were employed in the clean-up effort, including the use of detergents, absorbents, hot- and cold-water jetting, skimming, digging-up topsoil, etc. and some of these probably did more harm than good. In the initial years after the incident, residual oil contamination was observed to decline quite rapidly – between 50 to 65 percent per year – suggesting that any detrimental effects would rapidly dissipate. Work in subsequent years, however, showed a much slower rate of decline in residual oil, falling to around 4 percent per annum by 2003. As a result, some areas remain contaminated by oil from the spill and will remain so for some years to come. However, numerous studies also show that superficially, at least, the impact of the spill has largely disappeared.

In the case of the BP Macondo blowout, we can probably be even more hopeful, despite alarmist predictions that the Gulf of Mexico would be poisoned for years and that it would become a "dead zone". Thanks to warm surface waters and the use of approved chemical dispersants injected at depth into the rising oil column, much of the escaped oil had already dissipated within days of the successful surface kill operation that pumped heavy mud and cement into the well, according to the National Oceanic and Atmospheric Administration (NOAA). They estimated that almost 60% of the spilled oil had evaporated or been recovered and a significant percentage had already been biodegraded, thanks to the large indigenous population of oil-eating microbes resident in the GoM. They live and thrive in vast numbers there normally because of the natural oil seeps mentioned previously. In fact, while examining the seabed for leaks after the Macondo well was capped, natural seeps from a totally different oil reservoir were found a few kilometers from the damaged well, as if to illustrate the point. The media's disappointment was almost palpable and politicians railed at the NOAA for their irresponsibility in daring to suggest that things might get back to normal after all. What happened to all those beaches that were going to be oiled or the fabulous prospect of oil entering the GoM's Loop Current and contaminating the entire Eastern Seaboard?

Examination of the water column, in an area close to the Macondo well, by researchers from Woods Hole Oceanographic Institute, did however detect the presence of a sizeable oil contamination plume at a depth of 3000 to 3400 ft. Headlines proclaimed another apocalypse, painting a picture of monstrous toxic plumes of oil roaming beneath the surface of the now superficially-normal GoM. In fact, the oil in the plume was present at extremely low concentrations and the water had no discernible smell of hydrocarbons; samples were as clear as spring water, to quote the lead researcher. Given this degree of dispersion, it was very likely that this remaining oil would disappear soon enough and scientists said it was unlikely to cause any problem to commercial fishing or other marine life due to its depth. Finally, tropical storms and hurricanes in the Gulf of Mexico, far from being the problem that the media apparently wanted them to be, are nature's own way of dispersing and diluting spilled oil, just as happened in the *Braer* tanker incident already discussed.

In almost all of the incidents mentioned above, errors were made, and the consequences of those errors have been serious. However, as in every other field of human endeavor, lessons have been learned from such mistakes and steps taken to reduce the risk of their happening again, including legislation where appropriate. In the marine transportation sector, many countries now require double-hull tankers to reduce leaks in the event of collision, as discussed in Chapter 5. In the exploration and production business, when wells are designed and constructed in increasingly challenging frontiers, such as deep water, the pristine Arctic or inside urban areas, the well construction process is a sophisticated exercise involving many engineering disciplines. Although mistakes and accidents can happen, greater effort than ever is made today to ensure that the totality of the operation is safe and that such wells possess long-term integrity by applying the best isolation technology available. It goes without saying that regardless of the well location, the means must also be available, at short notice, to intervene effectively and curtail production in the event of catastrophe. Such lessons are often learned the hard way – not because of negligence or malfeasance but simply because nobody could imagine, a priori, the specific set of circumstances that ultimately led to disaster.

Society should recognize that, as long as we exploit this natural hydrocarbon resource for energy and use it to provide raw materials for our factories, the escape of oil will always constitute a risk. But it is a risk worth taking. Still, it is incumbent on the industry's professionals when planning, drilling, producing and transporting oil to minimize such risks and to find better ways to limit the environmental impact in the search for this valuable commodity. And, ultimately, since we all rely on oil and other hydrocarbons for the majority of our energy requirements and will continue to do so for most of the twenty-first century, this duty of care extends to all of us when we use them.

CHAPTER 11

Climate Change

Chapter 11
Climate Change

It is no exaggeration to say that energy is the most important resource of all. Nor to say that human beings love energy and will go to extraordinary lengths to ensure the supply of it. Why? Because harnessing energy freed mankind from humdrum tasks and allowed us to develop technologies and devices to accomplish things we would never have achieved, otherwise. And every year we develop more and more ways to use that energy to improve our way of life and to improve the efficiency of its use. Without energy, we are lost since all the things we take for granted in the modern world ultimately owe their existence or ready availability to our continuing access to cheap, reliable and adaptable sources of energy. Yet, perversely, the very resource that is intrinsic to our continued progress and development has become embroiled in perhaps the greatest public policy scandal of modern times - one that threatens the well-being of billions of people for generations to come. That scandal is the multi-billion dollar Anthropogenic Global Warming (AGW) business.

We should start this discussion by acknowledging that the Earth experienced a warming trend, natural or otherwise, in the latter part of the twentieth century. Such warming periods (as well as cooling periods) have occurred many times over the four billion year history of the planet and evidence of past events may be found in written and fossil records. However, helped by the mainstream media and militant environmental groups, this otherwise insignificant blip in the planetary temperature record has somehow become "proof" of an imminent global catastrophe. According to a theory with no single objective scientific proof to support it[54], the culprit for this impending doomsday scenario is carbon dioxide (CO_2), a trace atmospheric gas that is the source, ultimately, of the entire planet's food supply.

Yet that simple truth underlining the crucial role of CO_2 for every living thing on the planet has been overtaken by the unproven theory of AGW supported by an unlikely alliance of quasi-religious environmentalists, politicians, journalists, bankers and climate scientists. The environmentalists dream of returning to some mythical, pre-industrial Garden of Eden. The politicians want to be re-elected and make grand statements and future commitments that will be somebody else's problem to actually fulfill. The journalists just want a good story to sell their newspapers and have forgotten that their job is to report the news not create it. The bankers are looking for CO_2 to become the next "big thing" that will generate handsome profits and bonuses to replace dotcom start-ups and sub-prime mortgages. The climate scientists want funding and fame and recognize that both are possible by supporting the recent popular notion that CO_2 is a toxic pollutant. They also fear the ignominy and rejection of challenging the "scientific consensus" of the Intergovernmental Panel on Climate Change (IPCC), the UN group charged with reviewing climate science and reporting its conclusions to policymakers.

Leaked e-mails during 2009 from the University of East Anglia's Climate Research Unit, one of the most influential establishments in the study of AGW and contributors to the IPCC reports, show that this fear is well-founded.[55] The e-mails suggest that senior climate researchers modified data to eliminate trends that don't fit their models, denied other researchers access to their data (breaking the law in the process) and conspired to prevent skeptical scientists having their work included. The "*Climategate*" scandal, as it has been called, sent shock waves around the world just prior to the December 2009 Copenhagen Climate Conference and forced the director of the CRU to step down. He, and others in the AGW cabal, appear to have forgotten that science involves matching observation with theory and that if the real world data don't fit, the theory is almost certainly wrong.

The actual agendas of environmentalists, politicians, journalists, bankers and climate scientists may all be different but between them they have contrived to make CO_2, rather than energy, a valuable commodity that everyone else on the planet, will almost certainly have to pay for. In so doing, they are not just threatening our way of life but also condemning millions of people to poverty and almost certainly a shorter life span.

The Facts

In 2007, mankind helped send some 29 billion tons of carbon dioxide into the atmosphere[56], much of it from the burning of fossil fuels. This contributed to a trend, which has been ongoing since the start of the Industrial Revolution in the late eighteenth century, of a more or less continuous increase in the levels of atmospheric CO_2. Since 1832, atmospheric CO_2 levels have risen from around 280 ppm to 380 ppm. Fifty percent of that increase has been in the last 30 years, indicating an accelerating trend.

Despite the small absolute levels of the gas in the atmosphere, the relative change in concentration is viewed by AGW adherents, and, more recently, by politicians and portions of the general public, as significant and alarming. The main reason for this concern is that carbon dioxide is a so-called "greenhouse gas." Such gases are strong absorbers of infra-red (heat) energy and this phenomenon is responsible for warming of the atmosphere and the retention of radiative heat energy that would otherwise escape into space. The argument is that increases in average atmospheric temperatures are likely to cause increased evaporation of water from the oceans and, since water vapor is, in fact, a much more significant greenhouse gas than CO_2, given its concentration in the atmosphere, this may further exacerbate the problem. Rising temperatures may also cause melting of glacial and polar ice, causing sea levels to rise and causing loss of reflective surfaces (i.e., ice), again exacerbating the problem, and so on.

Climatologists, using computer models that attempt to integrate the many complex processes involved in determining global climate (or at least those we know about, or think we know about), have predicted that increasing levels of CO_2 will cause average temperature increases ranging from 1.4 to 5.8 °C (2.7 to 10.4 °F) between now and 2100.

Not everyone agrees with such assessments and there is a growing groundswell in the scientific community as well as amongst the general public that they need to challenge these models. There are also those among the scientists who agree that there is a possible ongoing warming trend (there has been no warming since 1998) but who are not convinced that the principal cause is anthropogenic CO_2 from the burning of fossil fuels. They point to long-term uncertainties in climate and cyclical changes, like Ice Ages, that have occurred many times over the life of the planet, related to volcanism, changes in the Sun's output, perturbations in the Earth's orbit, etc. They also point out that atmospheric CO_2 concentrations have been very much higher at numerous points during our planet's history with no obvious correlation to inferred global temperatures.

For example, during the Ordovician Era (~440 million years ago), CO_2 levels were estimated to be around 4,000 ppm, ten times more than current levels, yet the world was in the grip of a prolonged ice age. Even the Vostok ice core data, which are often quoted as being definitive proof of a link between CO_2 levels and global warming by AGW fans, clearly show that temperatures rose several hundred years before atmospheric CO_2 levels did, supporting the contention that the rising CO_2 levels were caused by, not the cause of, the rising temperatures.[57] An increase in temperature reduces the solubility of CO_2 in the oceans, a well-documented fact, and this would logically lead to an increase in atmospheric CO_2. The implication is that the CO_2 level is a trailing indicator of the temperature rise and certainly not the cause of it.[58] In fact, the warmest year on record so far was 1998 and several key climatic indicators, predicted by global warming adherents, have not materialized.

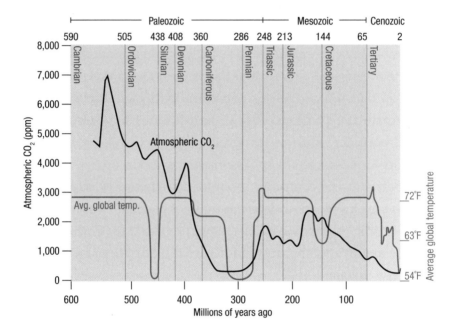

Figure 11.1. Historical CO_2 levels and global temperatures (Source: Geocraft.com)

Unfortunately, open scientific discussion and diversity in belief are not encouraged by the media or by the climate change movement. Perhaps they fear it may send a confusing message to the public and to politicians and so, in an effort to ensure a politically-uniform consensus, AGW proponents censure those who question the global warming dogma and label them as deniers, cranks, criminals or worse. The almost fanatical, quasi-religious beliefs that pervade the AGW movement are disturbing and, as noted previously, numerous serious scientists have been effectively ex-communicated by climate change adherents, funding bodies, government and academic institutions, simply for raising concerns about these matters. Some have received personal threats or are vilified in the press. Conspiracies to block their publications in scientific journals have been uncovered.[59] This can hardly be viewed as a constructive environment in which to conduct serious research or discuss alternative interpretations of data that challenge the established dogma.

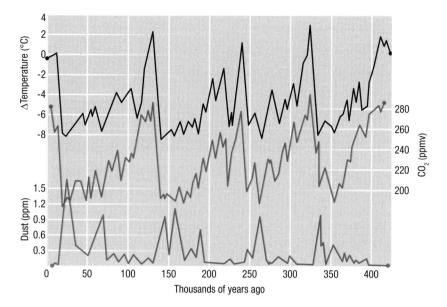

Figure 11.2. Data obtained from the Vostok ice cores showing fluctuations in atmospheric carbon dioxide levels and inferred global temperatures along the same time line. (Source: IPCC)

Like small ships driven on a stormy sea of public concern whipped up by media catastrophists and environmental crusaders, politicians, law makers and industry have been forced to take note and respond accordingly. They do so on the tenuous basis that it would be better to do something (or to be seen to be doing something) to avert impending doom, than to wait until it is too late – the so-called precautionary principle. Of course, in politics particularly, it is easy to gain short term kudos by making grand pronouncements of intent while knowing full well that some future government will need to deal with the reality of such long term commitments. As a result, huge amounts of public money have been thrown at the problem and an entire multi-billion dollar anti-global warming industry has sprung up to take advantage of this. The US government alone has spent almost $78 billion of taxpayer's money on climate research and climate-related technologies over the past 20 years with no empirical evidence that man made CO_2 has a significant effect on climate.[60]

It is true that not all of this has been a wasted effort and among other things it has stimulated research into alternative, reliable energy sources which has, in turn, created multiple business opportunities and spawned all manner of technical innovations. Scientists, business leaders, and engineers have scrambled to improve fuel efficiency, set ambitious goals for their organizations, reduce emissions or capture and "sequester" CO_2 before it enters the atmosphere. Less encouragingly, politicians, economists and legislators have dreamed up schemes to tax those who emit CO_2 and fine those who refuse to curtail their emissions. Unfortunately, since hydrocarbons, the source of much of that CO_2, provide more than 85% of our energy[61] and will for a very long time to come, this means that we will all have to pay, whether we like it or not.

The Science of AGW

At this point, it is worth pausing to review the basics of Anthropogenic Global Warming (AGW), the science that supposedly underpins it and the global business that has been spawned by it, discussing the key technologies and financial instruments in play and the implications of these on economic growth. We also review whether there are better and smarter ways to address this hypothetical AGW and its implications than by simply reducing CO_2 release to the atmosphere. And finally, we consider whether there may be other reasons why AGW has suddenly become the most important issue for politicians, environmentalists and certain other interested parties around the world.

It was recognized, as long ago as 1824, by the French scientist Joseph Fourier that gases in the earth's atmosphere might act to trap heat from the sun. But it was not until John Tyndall, an Englishman, conducted experiments in 1859 that water vapor and carbon dioxide were identified as the main constituents of normal air responsible for such an effect. In 1958, almost a century later, Charles Keeling, a scientist working for the Scripps Institute of Oceanography in La Jolla, California began a project involving the accurate measurement of atmospheric carbon dioxide levels. Keeling had previously been involved in the development of the first instrument to make accurate, routine measurements of atmospheric CO_2, which was present in normal atmospheric air at concentrations of only around 0.03%.

The site for this work was at an altitude of about 10,000 feet (3,000 meters), on Mauna Loa, a volcano in Hawaii. Keeling's work there, which has continued almost uninterrupted since its inception, showed a seasonal fluctuation in atmospheric CO_2 levels (due to consumption of the gas by growing plants in summer and release by decaying biomass in fall and winter), superimposed on a trend of a gradually rising baseline level of the gas. The early observations and recognition of an upward trend were confirmation of what had been suspected for over half a century – that burning fossil fuels for industry and transport was causing an increase in atmospheric CO_2.

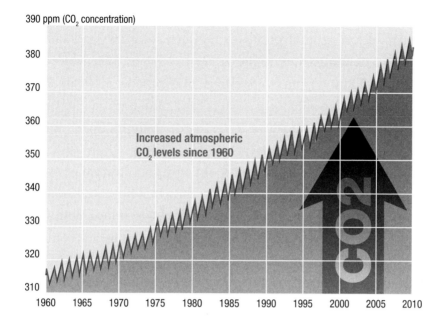

390 ppm (CO$_2$ concentration)

Increased atmospheric
CO$_2$ levels since 1960

*Figure 11.3. The Keeling Curve – Atmospheric CO$_2$ levels since around 1960
as measured on Mauna Loa in Hawaii. The rising trend is causing concern in
relation to increasing global temperatures.*

The individual usually credited with the observation that fluctuations in the level of atmospheric CO$_2$ might have important effects on climate was Swedish scientist Svante Arhennius, who published calculations in 1896 showing that a reduction in CO$_2$ could have been responsible for the prehistoric ice ages. He also calculated that doubling atmospheric CO$_2$ could result in a global temperature rise of 5-6 °C (9-11 deg °F). While his calculations are now known to have been over-simplistic, they are still remarkable for their prescience. Based on fossil fuel usage at the time, Arhennius estimated that it would take 3,000 years for atmospheric CO$_2$ to double, rather than the one or two centuries, based on current projections. He also viewed the possibility of a global warming trend as a positive development, likely to improve crop yields and to reduce the risk of the world slipping into another ice age. One century on, the fear being spread by AGW alarmists is not of another ice age but of a planet with temperatures too hot to sustain our way of life. However, there is much more to this than meets the eye and, while we cannot cover all aspects of this issue, we will try to put it in proper context and discuss some of the key scientific and political ramifications of these matters.

The Implications

Most members of the general public in modern societies have only a sketchy idea of the complex infrastructure necessary to sustain their way of life. Stores are full of goods and food from every corner of the planet is cheap and plentiful. Power is available at the flick of a switch, and transport, at the turn of a key. Many view this as normal and give it very little thought. Some may marvel at the convenience and choice afforded by modern living while still others would rather live some romantic version of the old agrarian lifestyle of our forebears. Yet few would wish to give up the many benefits that have accumulated during mankind's rapid progress from field to factory, hut to house, commune to city. Improvements in sanitation, health, diet, life expectancy, literacy, communication, transportation and living standards, reductions in child mortality, famine and disease, the development of medicines, plastics, pesticides, fertilizers, fabrics and industrial materials, the invention of radio, television, computers, aircraft and motor vehicles: all of these, and many, many more, have had an enormous impact on the quality of life of vast numbers of people.

It is easy to say that we must make changes in the way we burn fossil fuels so that emissions of CO_2 into the atmosphere can be reduced. However, it is much more difficult to implement those changes when they will directly affect the lifestyles to which many people are accustomed and when no practical alternatives exist, in the foreseeable future. No more gasoline-fueled vehicles, no air travel for vacation, no electricity, no fertilizer to grow crops or insecticides to protect them, no plastics, no fuel for cooking, no goods to buy in the stores, no international trade, etc. These might seem extreme scenarios but they are what would happen if the world suddenly decided to stop using fossil fuels overnight.

This, of course, will not happen and, despite all the rhetoric of politicians and environmental groups, it cannot happen. The world will continue to be powered predominantly by hydrocarbon fuels at least through the next 50 years. What happens beyond that timeframe will depend on a number of factors, some technical, some political and some financial. Without doubt, however, coloring the whole picture, certainly for the next few years, will be the issue of global warming.

As more data are gathered, hopefully real science, rather than quasi-religious "End of Days" beliefs, will establish how our climate is changing and what steps we need to take, if any, to mitigate any negative effects of that on our civilization. In the meantime, the two biggest "Grand Policy" ideas to constrain the inexorable rise in atmospheric CO_2 concentrations will no doubt be implemented in some form in different countries, regardless of whether there is actually any need to do so from the perspective of the planet's future wellbeing. These two strategies – to capture emissions and store them or to levy some form of taxation to curb them – are discussed next.

The Present Carbon Cycle

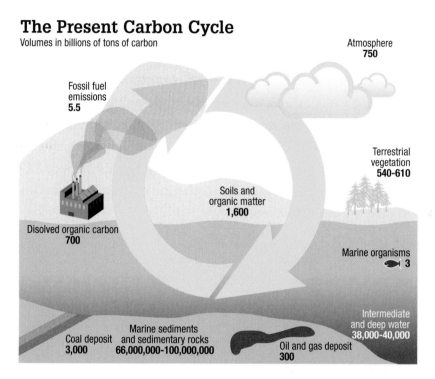

Figure 11.4. The global carbon cycle as it is largely understood today. Estimates vary somewhat but certainly the largest fluxes of carbon dioxide are between the atmosphere and the oceans which act as a partial sink for CO_2. (Source: UNEP)

Carbon Capture

We have mentioned already in Chapters 1 and 8 that the biggest contributor to CO_2 emissions today is the power generation industry, accounting for 40% of the world total. Since the power is typically generated at large scale coal-burning plants, these can be considered as focal points in the climate change issue. A typical 1,000 MW coal-burning plant releases some 8 million tons of CO_2 per year, equivalent to that produced by one million cars powered with conventional gasoline engines. It is certainly easier and cheaper to remove the CO_2 at a central facility than it is to fit removal equipment to a million vehicles. So, it would seem to be a simple enough proposition – remove the CO_2 from the power plant waste stream and we can cut emissions significantly.

However, the capture of CO_2 requires special equipment and incurs additional cost that would need to be paid for by consumers. Also, separating the CO_2 is only part of the problem. What do we do with the CO_2 once we have captured it? There are some industries that could use some of the CO_2. For example, it could be used to increase growth rates in greenhouses or in algae farms producing biodiesel. However, this would consume only a small amount of the available gas, so a long term disposal (or storage) repository is required for the bulk of the CO_2 and, in theory, the most logical place to put it is deep underground. It can be injected into old oil and gas reservoirs or into other suitable rock layers with acceptable vertical containment barriers (e.g. cap rock) to prevent leakage. Superficially, this would appear to be a workable solution but, of course, it would require every power station to have dedicated disposal wells and suitable underlying geology. A quick look at the power station map in Figure 8.6, gives an idea of the magnitude of that task without even considering the geological criteria.

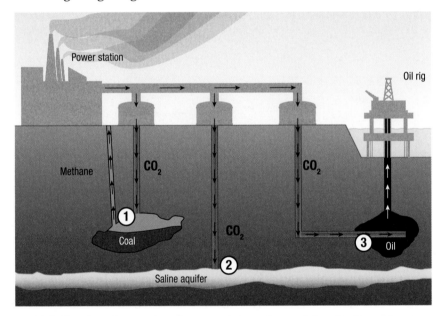

Figure 11.5. Basic schematic of processes envisaged for Carbon Capture and Storage (CCS). Potential repositories must be carefully selected to ensure long-term storage with minimal leakage. (Source: World Coal Institute)

Carbon capture and storage or sequestration (CCS) has been the great hope of middle-of-the-road characters, all those between environmental extremists ("greenies") and their "right-wing" adversaries. Governments and certain oil companies have taken to CCS and it has figured prominently in many energy and climate scenarios by think-tanks and universities. In some ways it has acted as an "indulgence" certificate. The logic goes that no matter what one's position is on AGW, we have a way to paper over the debate. Engineers, and in particular, petroleum engineers, have a solution – bury the problem by re-injecting the offending gas back into the ground.[62] Targets can be old oil and gas reservoirs but, more mentioned, are deep saline aquifers of which there are plenty.

Several oil companies tout CCS and some petroleum engineers, unfortunately avoiding simple calculations, have jumped on this bandwagon. The stakes are high and the rewards for those promoting the idea are lucrative. Government and even company funding is plentiful for this. And as usual, many researchers, instead of studying the very feasibility of the process, already taken for granted in some circles, have started working on peripheral aspects, such as solubility of CO_2 in water and even mineralization – processes that take tens to hundreds of thousands of years to mature.

The reality is very different. What have rarely been calculated, although talked about, are the rate of CO_2 injection per well and the cumulative volume of injection in a particular geologic formation. These are critical elements of the process. In a paper published in October 2009 the feasibility of sequestering CO_2 as a means of emissions management was addressed.[63] The conclusions are quite sobering. The calculations presented suggest that the volume of CO_2 to be disposed cannot exceed more than about 1% of the pore space of the geologic formation. This will require from 5 to 20 times more underground reservoir volume than has been envisioned by many, and it renders geologic sequestration of CO_2 a profoundly non-feasible option for the management of CO_2 emissions. Kyoto Protocol or successor accords would imply a problem that is orders of magnitude larger than anything possible as CCS.

Note too that published CO_2 injection rates based on Enhanced Oil Recovery (EOR) experiences, assuming open underground aquifer conditions, fail to reconcile the fundamental difference between steady state, where the injection rate is constant, and pseudo-steady state where the injection rate will undergo exponential decline if the injection pressure exceeds an allowable value. A limited aquifer realistically indicates a far larger number of required injection wells for a given mass of CO_2 to be sequestered and/or a far larger reservoir volume. The implications are profound because simple models assuming steady state injection conditions are missing the critical point that the reservoir pressure will build up under injection at constant rate. Instead of the 1% to 4% of bulk volume storability factor indicated prominently in the literature, a more realistic finding is that CO_2 can occupy no more than 1% of the pore volume and likely as much as 100 times less.

Figure 11.6. Lake Nyos in Cameroon shown after the spontaneous out-gassing of volcanic gases (CO_2 & H_2S) from the lake in 1986

The volume of the reservoir that would be adequate to store CO_2 is very much related with the need to sustain injectivity. The two are intimately connected. In applying this to a commercial power plant of just 500 MW the findings suggest that for a small number of wells the extent of the reservoir would have to be enormous, the size of a small US state. Conversely, for more moderate size reservoirs, but still the size of Alaska's Prudhoe Bay reservoir, there would be a need for hundreds of wells. Neither of these bode well for geological CO_2 sequestration and the findings of that work clearly suggest that it is not a practical means to provide any substantive reduction in CO_2 emissions.

In addition, and apart from the long-term storage problem, there is also the risk of abrupt, unforeseen leakage from these repositories. Carbon dioxide is an asphyxiant and, being denser than air, in an enclosed space or low lying area it can suffocate people by displacing air. Such a scenario unfolded several years ago at Lake Nyos in Cameroon when a volcanic disturbance caused the release of an estimated quarter of a million tons of naturally-produced CO_2 (mixed with some hydrogen sulfide (H_2S), a poisonous gas found around volcanoes) from the lake bottom, where it had collected through natural processes.[64] The cloud of CO_2 and H_2S poured down the slopes of the volcano and into two nearby valleys killing 1700 villagers and thousands of cattle.

Taxing Carbon

Given that the majority of CO_2 we have been discussing arises from the consumption and combustion of fossil fuels and that the majority of that consumption is in the developed world, an obvious solution is to make people there pay for each ton of CO_2 produced. The mechanisms to do this vary from a simple increase in fuel tax, by imposing a nominal "Carbon Tax" (on gasoline, coal and electricity) to more complicated models involving carbon credits and offsets, the so-called "cap and trade" approach.

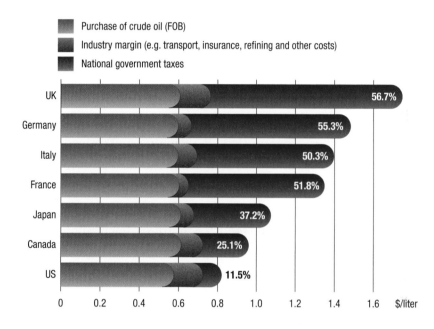

Figure 11.7. Motor fuel taxes in the G7 countries. Energy taxes are common in most rich countries and can represent a significant percentage of retail price. While providing valuable income to the respective governments such taxes also increase costs across the entire economy. (Source: EIA)

Taxes are nothing new, of course; nor are taxes on fuel. In some shape or form, taxes on various forms of energy have been around for a long time. Taxes on candles, even windows and thus daylight, were introduced in eighteenth century Britain and taxes on coal, gasoline and other fuels all appeared during the twentieth century. Many countries have an electricity tax and, in general, this tax is levied at the same rate, regardless of the provenance of the electricity consumed. Thus, governments can raise significant revenue from such direct taxation of fuel and energy. It should be noted, however, that such taxes affect the poor disproportionately.

Fuel and energy taxes increase the cost of all materials, since transportation costs are a component of virtually every product on any store shelf. They also have a direct impact on the cost of private and public transportation and its availability while raising bills for electricity and every other utility that uses energy, in any form. However, direct taxes on fuels have the benefit of simplicity and transparency, if nothing else. They are directly related to the commodity being consumed (oil, gas, electricity) and there is a clear relationship between consumption and cost.

The main alternative regime to such direct taxation, so-called cap-and-trade, is much more opaque and, attracts criticism from many quarters. However, in essence, cap-and-trade is relatively simple and requires just a few minimum features.[65]

- **An "Emissions Cap"**: This sets a mandatory limit on the total number of tons of the pollutant that can be emitted. It provides a standard for measuring environmental progress, and it provides a genuine value for those tons of pollutant traded on the markets.

- **A fixed number of allowances for each emitter:** Each allowance gives the owner the right to emit one unit of pollution at any time. The unit is normally measured in tons. The methodology used to allocate allowances may vary with time, based on policy requirements, etc.

- **The opportunity for banking and trading.** An emitter that reduces its emissions below its allowance level (e.g. by investing in new technology or changing processes) may sell the surplus allowances to another emitter and make a profit. An emitter that is stuck with old equipment and methods may purchase allowances from another emitter to stay in business. Buyers and sellers are allowed to "bank" any unused allowances for future use.

- **Clear performance criteria.** At the end of the compliance period, each source must hold a number of allowances equal to its tons of emissions for that period, and must have measured its emissions accurately and reported them transparently.

- **Flexibility.** Sources have flexibility to decide when, where and how to reduce emissions.

The Clean Air Act in 1990 formed the basis for such cap-and-trade systems for sulfur, whose oxides, as by-products of coal-burning power plants, form sulfuric acid in contact with water and air and were widely blamed for causing "acid rain," an insidious form of pollution that caused widespread damage to forests and lakes in northern latitudes during the 1980s.

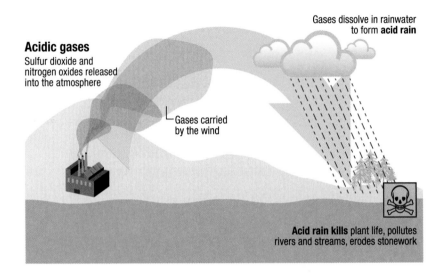

Figure 11.8. Schematic of the creation of acid rain due to SO_x and NO_x produced in coal burning power plants. (Source: IPCC)

The act set a so-called "cap" on emissions, limiting the absolute amount of sulfur dioxide that could be vented to the atmosphere to a level 10 million tons below the levels vented in 1980. While threatening penalties to those who exceeded their emission quota, the Act also provided incentives to those who found ways to reduce emissions (e.g. by using new technology). This approach allowed others to pay for their continuing emissions by buying offsets (credits) from others who no longer needed their full quota. This ability to trade virtual credits creates a market for these credits and the market should, in principle, converge on an acceptable price range for such credits, balancing the various forces in play between the seller of a credit and those who want to buy it. Such an approach rewards those who quickly adopt new technologies but also allows those who cannot make rapid changes to their processes or facilities time to adapt whilst continuing to operate as a business. In the longer term, however, planned reductions in the cap make it progressively more expensive and difficult to meet the emission standards profitably, forcing older, inefficient plants to modernize or close. A final and key component of cap and trade is the global interchangeability of trading credits. Thus, while gradually capping emissions in the main producing countries it provides tradable credits for those (usually less developed) countries whose emissions are below their current cap. These can be sold at market value and so provide a source of additional income to poorer countries.

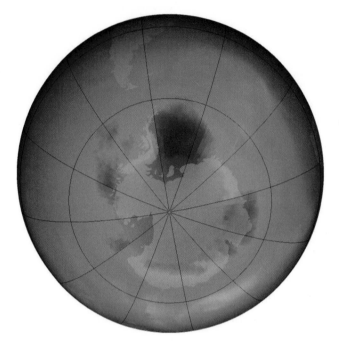

Figure 11.9. Satellite image showing the Antarctic "Ozone Hole" created in the atmosphere due to the interaction between sunlight and CFCs in the upper levels of the atmosphere.

The principles behind cap-and-trade for CO_2 are the same as those used for other previously defined pollutants, like sulfur dioxide (SO_2), among the oxides of sulfur (SO_x). At first glance, cap-and-trade would seem to be a clever mechanism to deal with CO_2 emissions. It has all the elements for such an approach and the experience gleaned from control of sulfur emissions and production of chlorofluorocarbons (CFCs) would prove invaluable. (CFCs have been identified as the main cause of the loss of ozone, a form of oxygen, from the upper levels of the earth's atmosphere. Ozone is important because it shields the earth's surface from dangerous levels of ultraviolet radiation that would otherwise reach us from the sun. Agreements signed in Montreal in 1989, limited the production and release of ozone-depleting CFCs).

The similarities between a cap-and-trade approach for CO_2 and one for SO_x/CFCs, however, are dwarfed by the differences. Nobody doubts that the release of sulfur dioxide or CFCs to the atmosphere is a bad thing. The products are clearly harmful pollutants and technologies exist to reduce their use (e.g., by replacing CFCs with other, less harmful refrigerants and propellants) and release (e.g., by scrubbing flue gases from power stations to remove sulfur). But no such simple and cheap replacement or removal technology exists for CO_2.

Unfortunately, no matter how we slice it or dice it, extracting energy from hydrocarbons invariably involves burning (or oxidizing) them and that inevitably leads to the production of CO_2, along with the energy. CO_2 is much more difficult and costly to remove than SO_x and, in effluent from a power station, for example, is present at concentrations many orders of magnitude greater than those of its sulfurous relatives.[66] Accordingly, the cost of removal (and eventual disposal) of these much greater volumes of CO_2 is much, much higher.

However, the most important thing to note here is the classification of these very different materials – all three are now officially classed by the EPA as pollutants. This begs the question, how can CO_2, the principal product of plant and human respiration, one of the two main constituents of the entire food supply for the planet (the other being water, both required for plant photosynthesis which, ultimately, provides all food on our world[67]) and the inevitable by-product of our dominant energy supply for the past couple of centuries, be classified as a pollutant? This is, quite simply, nonsense.

But now that CO_2 has been thus classified its emissions can be capped and traded just like other pollutants. The sheer scale of that trade, however, and its potential market value, easily overshadows the current market for earlier pollutants. As noted above, the total market for SO_x is in the range of a few million tons while that for CO_2 is over a thousand times bigger. Thus, a tax on CO_2 whether direct or otherwise, would raise billions of tax dollars for governments. Conveniently, if instituted quickly, it would do so during a period when many governments' budgets are in dire straits due to the recent financial crisis and the need to bail out banks and other institutions. Even better for government, however, is the likelihood that a carbon tax can be raised on an ongoing basis, year-on-year, with the possibility of increases in line with market conditions and the ever-tightening cap on total emissions. Such multi-trillion dollar tax opportunities rarely present themselves to governments and it is almost certain, therefore, that, regardless of the validity of claims that our burning of fossil fuels is affecting the climate, we will see increasing taxes on traditional energy sources, in the very near future. How else will governments rebuild their damaged financial positions and provide a vast pool of money with which to fund any number of government programs, many, no doubt, totally unrelated to the problem for which the money was supposedly raised in the first place.

On the other side of the equation, poorer countries which may be at risk from climate change stand to benefit to the tune of billions of dollars in additional aid from rich countries. Free money is always a dangerous thing unless its disbursement is carefully controlled and is targeted towards the specific use for which it was intended. The fact that fiscal control and accountability are widely acknowledged to be problems already in many of the potential recipient countries suggests that aid may not always be invested wisely and may instead shore up corrupt or bankrupt regimes.

Other technical issues complicate cap-and-trade on CO_2 and potentially make this option more open to abuse and manipulation. As noted already, CO_2 is not the only greenhouse gas. Water vapor is a major contributor to the greenhouse effect and some other more exotic molecules are more potent still (HFC-23, a by-product in the manufacture of refrigerants, is some 11,700 times more potent than CO_2). In the UN's Clean Development Mechanism and the EU's carbon trading scheme, these other chemicals are considered to be greenhouse gases and strategies to limit their production and release to the atmosphere are eligible for carbon credits, on a ton-for-ton basis in line with their greenhouse gas tendencies, referenced to CO_2 as the de facto standard "greenhouse gas." Thus, if a factory in China destroys, say, 200 tons of a highly potent greenhouse gas byproduct like HFC-23, it is potentially eligible for carbon credits equivalent to around 2.5 million tons of CO_2. It can now sell these same credits, at their current market price, to companies in the industrialized world who need to continue burning fossil fuel and no longer have sufficient permits to do so (due to the tightening emission cap).

The potential for such schemes to distort markets and to enrich clever unscrupulous operators is obvious. In the above example, the factory could elect to increase refrigerant production simply to increase its production of the highly lucrative byproduct, which it then destroys and, in so doing, makes more profit than it generates from the refrigerant product itself. However, not one ton of actual carbon dioxide has been prevented from entering the atmosphere. Instead, it is likely that more has been generated at higher cost to society and significant gain to a few clever people.[68]

Even in those cases where carbon credits are "real," insofar as they represent genuine reductions in CO_2, what exactly are we paying for and who benefits from this transaction? In truth, we are paying for nothing. Worse still, because these credits can be traded, we are creating a totally fictitious market vulnerable to speculation and all the worst excesses that we have already seen in equity, credit and derivative markets in recent years. Referring to financial derivatives in 2002, Warren Buffet, the well-known investment guru, famously called them "financial weapons of mass destruction" due to their tendency to mask accounting imbalances and their ability to elicit huge market effects in response to relatively minor events. The carbon market is conceivably the biggest derivative market of them all, trading in a totally worthless virtual commodity with no intrinsic value, the price of which is subject to speculation and manipulation. Such speculation could dramatically increase the cost of every real product and service in the market and ultimately affect the economic output of the entire world. The impact of a speculative carbon price bubble would be much worse than an oil price hike for the simple reason that it would affect 85% of our energy supply and have knock-on effects on everything else.

The carbon trading market was estimated to be worth $136 billion in 2009 up from a mere $58 billion in 2007. That $136 billion represented 8.2 billion tons of carbon dioxide, according to consulting firm Point Carbon, or slightly over a quarter of the entire world's total CO_2 emissions. Analysts in the carbon trading business project that it will grow to become a $2 trillion to $10 trillion market. Huh? How does that work?

In 2010, the market price for CO_2 is around $17 per ton, based on the figures above. With total global emissions of CO_2 in the near future at, say, 35 billion tons, how can the market ever be worth $10 trillion dollars? Such figures imply a CO_2 price of $300 per ton or about twenty times higher than today's supposed CO_2 price. That's the equivalent of paying an extra $2500 a year for gasoline for your car or an additional tariff of 30 cents per kWh on electricity. The impact of such extra and totally unnecessary costs on world economic development will be crippling. They will reduce GDP across the board and will inhibit investment and research in areas where they are genuinely needed. At the same time they suck money out of the real economy, these measures will be generating obscene profits for banks and financial institutions on the basis of nothing more than hot air.

And what else could that money have been spent on? A tiny fraction of these trillions of dollars could help the fight to eliminate poverty and improve health care. How about just buying bed-nets to try to prevent a million people dying every year from malaria? Or how about investing in sustainable agriculture or infrastructure projects to help the developing world to provide electricity to over a billion and a half people who have none to this day? Or investing in research on cancer or HIV or a multitude of ailments that afflict all of mankind? And, of course, for the tiniest fraction of this money wasted on a dubious war against an unsubstantiated enemy, we could help poorer countries protect their people, their coastlines and their economies against the possible, but certainly addressable, challenges that may yet arise from natural climate variation of the sort that is most likely in effect now.

CHAPTER 12

Energy Future

Chapter 12
Energy Future

We have made a case in this book that energy and energy supply are essential to the world as we know it and that, while energy conservation and efficiency are desirable and indeed necessary, they will not reduce energy demand in the foreseeable future, if ever. Both the developing world in trying to catch up and the developed world, in finding new uses of energy, will see to it.

It is then instructive to attempt to forecast energy future and in the beginning of the twenty-first century there are two scenarios that one may envision. The first is a carbon constrained world, based on the relatively recent attempts to link man-made CO_2 with global climate change; the second is unrestricted use of all potential energy sources as dictated by market forces and the physics of the individual sources.

Irrespective of the authors' own views, one must certainly consider both notions, even if recent events and public opinion may have tilted the balance inexorably against the carbon-constrained future.

Before we attempt to analyze the various potentials it is essential that we state up front certain important premises:

1. Government decisions, by definition political, may generate situations that are neither natural nor are they supported by market forces.
2. Subsidies may generate entire industries which may appear healthy but can readily collapse once the subsidies disappear or are reduced.
3. Manufacturers of equipment may tout their environmental orthodoxy by saturating the media with advertisements, creating public opinion in favor of a particular energy source, creating a reality that is not. Lobbying government and legislatures further promotes a make-believe situation that is based on very flimsy rationale.
4. Technology of course will also play a pivotal role and certain seminal technologies, properly and strategically deployed, can become game changers. Two examples of technology which have either already made a difference en route to much bigger effect or on the way to dominance will be presented at the end of this chapter.

5. It becomes difficult to divorce myth from reality, fiction from fact, and it makes little difference whether the proponents are outright deliberate liars or whether they and their supporters, believe and propagate their own myths which may be consistent with a wishful thinking and a fantasy on how the world should be.

Let's offer two examples:

Relative sizes of alternative energy physical plants

Many equipment manufacturers, engineering firms, power companies and even oil, gas and coal companies, to boost their environmental credentials/images, publish photos of conventional power plants ("the past," at times complete with a smoke stack, which may actually be a steam vent, along with the massive venturi coolers.) The next photo almost invariably contains three to five wind turbines ("the future" in a pristine setting with green grass surrounding the virginal white of the turbines.) There is a serious problem with this not so intellectually honest depiction. A modern coal fired power plant may have a capacity of 2,000 MW. With a capacity factor of 75% or higher it means that it can deliver on average 1,500 MW into the grid. Now consider wind turbines and their load factors. While some very optimistic statements have suggested 25% or more, actual experience shows the capacity factor to be 15% or less. This means that 10,000 MW of wind turbines must be installed to deliver 1500 MW into the grid. With 1MW capacity wind turbines this would mean 10,000 turbines. With 2 acre spacing this would lead to 20,000 acres or more than 30 square miles. More than 4,000 of the much larger 2.5 MW turbines would be needed but the areal extent would probably not be reduced.

What this simple calculation suggests is that the image of the three turbines that to the uninitiated person may imply that they can replace the coal fired plant should be replaced by a huge forest of turbines covering the areas of several hills.

Relative magnitude of energy sources

Corn based ethanol and soybean biodiesel have been invariably offered as alternatives to petroleum derived motor vehicle gasoline and diesel. However, a simple calculation shows readily that if all the corn grown in the United States were to be converted to motor vehicle ethanol without regard to what that would do to food prices, it would amount to only about 20% of the US gasoline demand. If all soybeans were converted to biodiesel, that would be less than 4% of US diesel demand. Yet, in spite of this unassailable fact and its inherent problems, the most important of all being the impact on food prices and the questionable impact on global climate change, these biofuels have been unabashedly promoted as a replacement for petroleum liquid fuels and, in fact, the preferred mechanism towards "energy independence." With the United States importing almost 70% of its petroleum needs, corn based ethanol cannot even come close to fulfilling that goal, another clearly dishonest attempt toward energy solutions.

Energy in a Carbon Constrained World

In 2009, the US National Academies of Science and Engineering published an extensive report, entitled *America's Energy Future* (AEF), following considerable deliberations and research by dozens of well known scientists and engineers. Without once offering even remote homage or even mention of any skeptics and doubts on anthropogenic global warming (AGW) and without justifying or quantifying the impact of CO_2 on climate, the AEF report accepted not only the entire premise but also ancillary elements such as the alarmist implications of having to take drastic and immediate action, presumably to prevent something that others have labeled as the "tipping point."

Such attitude, suggesting trillions of dollars of national and international transformations, in the face of growing scientific objections and, especially massive public doubts on AGW and even more weariness on what to do about it, is mystifying. It proves that even scientists are not immune to the socio-phenomenon Charles Mackay noted that "Men go mad in herds and only recover their senses slowly and one by one."

The resulting report is astonishingly naïve and transparently biased. And in many ways it provides a disservice to the nation. We offer below a summary of its highlights, to allow the reader to extract his/her own conclusions, in spite of our reaction. It is instructive on the challenges to formulate a sane energy future.

The AEF report attempts a synthesis of expert opinions on the future of energy for the United States. The specific charge for the Report Committee was to *"critically evaluate the current and projected state of development of energy-supply, storage, and end-use technologies. The study will not make policy recommendations."* (p. 5) *"The committee will develop a 'reference scenario' that reflects a projection of current economic, technology cost and performance, and policy parameters into the future."* (p. 10)

One would expect that such a report would present a careful assessment of the availability of energy sources in the future, along with constraints imposed by technological, environmental and economic factors. However, in spite of the vow not to make policy recommendations the AEF report is loaded with a politically-motivated view of the world and future. Furthermore, what dominate the discussion exclusively are constraints imposed on our energy future by concerns about potential long-term effects of global climate change and anthropogenic greenhouse gases. To be sure, concerns about human impact on the environment are not unwarranted, and resulting action can have beneficial effects, as, for example, the Clean Air Act has shown. (p. 26) But the urgency and extent of action against global climate change proposed in the report may create more major problems than such action may solve, particularly economic problems, but also environmental. To be specific:

- It is ironic that the significance of a coordinated global effort against global climate change is relegated to the level of a brief footnote: (p. 68) *"Comparable actions at existing fossil-fuel plants in other countries will also be required to achieve substantial reductions in worldwide CO_2 emissions."* While pioneering related technology might benefit future technology exports for the US, the cost of developing and implementing such technology would be outrageous, and would prove an exercise in futility, were a global effort not to be undertaken. The debacle of the 2009 Copenhagen Climate Change conference illustrates that bold rhetoric is not always followed by bold acts, particularly when the cost is enormous and the outcome dubious.

- Urgency is emphasized throughout the Report, which is perplexing. A specific example: (p. 26, footnote) *"The committee refers in particular to uncertainties in the time-dependent relationships associated with anthropogenic CO_2 emissions and the resulting changes in atmospheric temperatures and sea levels. These uncertainties make it difficult to judge precisely how soon CO_2 emissions must be reduced to prevent major environmental impacts around the world. Many experts judge that there are at most a few decades to make these changes."*

  - According to a 2008 publication on climate change by the National Academies[69] (p. 9) **"by 2100** ... temperature *increase will be accompanied by an increase in global sea level of between **0.59 and 1.94 feet (0.18 and 0.59 meters)."***

    - First, this is a prediction for 90 years from now. How certain can one be over that period?

- Second, assuming the prediction is valid, is a sea level rise of 1.94 feet (sic) cause for alarm? Can one honestly claim that the potential damage from, say, an earthquake in San Francisco is lower than the potential damage from two feet of sea water? Would Noah panic?

  - In defiance to climate model predictions, global temperatures have not risen in the last decade, despite the fact that CO_2 emission increases have continued unabated. How much, then, should one trust such predictions, and, indeed, the very premise of the magnitude of the effect CO_2 has on global climate as a greenhouse gas (GHG)?

  - For changes that will take decades to materialize, missing a year or two may not be all that crucial, and could even be beneficial as additional information accumulates and mistakes in locking in potentially unsuccessful solutions are mitigated. For example, thermonuclear (hot) fusion has cost billions of dollars and has been constantly thirty years away from commercialization for the last fifty years. In fact, it appears that hot fusion will not ever work.[70] Would it have made a difference if development efforts had started a year earlier or later?

Despite the fact that technology cost assessment is in the Committee's charge, a clear reference to the cost of carbon capture and storage (CCS) appears only towards the very end of the Report: (p. 105) *"Retrofitting for 90 percent CO_2 capture at existing PC plants … would require capital expenditures **approaching those of the original plant itself**; and **20–40 percent of the plant's energy** would be diverted for separation, compression, and transmission of the CO_2, thereby significantly reducing thermal efficiency and increasing the levelized cost of electricity."* Yet, even though CCS would substantially increase the cost of electricity, it is portrayed in the Report as panacea, and is recommended (p. 5) as one of the two technologies to test immediately by 2020 (the other being evolutionary nuclear power). Of course, carbon capture is not the only component of CCS. Sequestration, which the AEF does not even tackle, is a far more cumbersome issue. And yet CCS is a presumption for the AEF report to allow continued use of fossil fuels from which 85% of primary energy is derived. Regarding nuclear technologies (p. 5), the spent fuel issue is not discussed until towards the end of the Report (p. 114).

"High-priority technology demonstration opportunities during the next decade include… cellulosic ethanol." (p. 6). Is it not the case that bio-fuels suffer from the inherent limitation that photosynthesis (the process which converts solar energy

to biochemical energy in plants) has only 1 to 3% efficiency[71], hence sustainable energy from biofuels in significant amounts is rather unrealistic, without major changes in land use and farming practices.

It is also ironic that within less than a year from the publication of the AEF Report, significant changes have already taken place, e.g.:

- (p. 4) *"…hydrogen fuel-cell vehicles."* DOE under Steven Chu cancelled the Hydrogen Car program[72] (but maintained the fundamental research program on fuel cells – both correctly in our opinion).
- US recoverable natural gas reserves have experienced an enormous increase, due to recoverability of unconventional resources.

The AEF report reviews the status and anticipated developments of energy-supply and end-use technologies. The report admits that fossil fuels contribution will change little in the near future. The crucial role of petroleum as practically the only energy source for transportation is emphasized. The debate about peak oil and the timing of the eventual depletion of petroleum is touched only in passing (in footnote 7, p. 14).

But then the report reveals its biases. It is argued that the main challenge for the US is to develop *"new technologies that use energy more efficiently and that avoid, or capture and safely store, greenhouse gas emissions"* and that *"failure to develop and implement such technologies will greatly limit the options available for reducing the nation's greenhouse gas emissions to the atmosphere."*

And it gets worse:

- p. 10: While the study charge list states *"Key environmental (including CO_2 mitigation) impacts,"* the Report focuses almost singularly on CO_2, while downplaying other environmental issues, e.g., the effect of new technologies on water resources. To wit: (p. 11) *"The burning of fossil fuels has a number of deleterious environmental impacts, among the most serious of which is the emission of greenhouse gases, primarily carbon dioxide (CO_2)."* Yet, in the same page it is stated: (in footnote 11) *"… uncertainties … associated with anthropogenic CO_2 emissions and the resulting changes in atmospheric temperatures and sea levels … make it difficult to judge precisely how soon CO_2 emissions must be reduced …"*

• p. 11: *"And despite improvements in energy efficiency, U.S. energy consumption continues to rise"* should read *"…"despite,* and often because of *improvements in energy efficiency…"* It is a well proven fact that once energy efficiency makes an activity (e.g., air conditioning, driving a second car) inexpensive enough for an individual or family, then that activity, which would otherwise be avoided, will now be undertaken, thus adding to the total energy consumption.

The AEF Committee considered technology development and deployment over three time periods – 2008–2020, 2020–2035, and 2035–2050 – but focused mainly on the first two periods, because of increasing uncertainty in projected future scenarios.

"… A major message of this report is that the nation can achieve the necessary and timely transformation of its energy system only if it embarks on an accelerated and sustained level of technology development, demonstration, and deployment along several parallel paths between now and 2020."

Three panels prepared reports on: energy efficiency, liquid fuels for transportation and electricity from renewables. The Committee chose not to address

• The future role of technologies for the exploration, extraction, storage, and transportation of primary energy sources (e.g., fossil fuels);
• Energy systems at the regional level;
• How various technologies might compete in the marketplace;
• The relative desirability of energy supply options;
• Detailed cost estimates (as opposed to rough cost estimates, which are addressed);
• The full range of options for reducing energy use (via efficiency and conservation)
• Future forecasts for energy prices.

It is a fatal flaw of this report that future technologies for fossil fuels are deliberately left out. The reason cited is: *"The focus of the report is on energy-supply and end-use technologies that are most likely, in the judgment of the committee, to have meaningful impacts on the US energy system during…the next 40 years or so. However, the committee did not assess the future role of technologies for the exploration, extraction, storage, and transportation of primary energy sources (e.g., fossil fuels), nor did it assess the role of some critical components of a modernized infrastructure—*

including tankers, roads, pipelines, and associated storage facilities — in delivering these resources from suppliers to consumers."

Yet in the same Report, the role of fossil fuels is not questioned, in any future scenario. Incremental improvements or breakthroughs could have significant implications for fossil fuels. To wit, recent successes in extracting unconventional natural gas have had a significant effect on the recoverable reserves of the US. And a breakthrough on, say, natural gas hydrates would have a tremendous impact on the energy landscape.

We believe that the choice made by the AEF Committee on fossil fuels, along with their fixation on CO_2 emissions and belief in CCS obfuscate future energy prospects for the US, and find that the report only does a disservice both to the scientific community and to the nation.

A More Logical Forecast and a Proposal

It is our belief that a carbon constrained world will simply not materialize, and not even remotely to the degree that the last few years of the first decade of the twentieth century might have signaled through political proclamations or even studies such as the one by the US Academies of Science and Engineering. There are two simple reasons for this. First, the presumed science and the alarmism it has spawned simply will not be borne by the facts and actual climate data. Evidence is already amassing and the public, as poll after poll suggests, is neither convinced nor does it consider climate change and its remedies as a national priority. Second, important countries such as China and India will never agree to mandatory carbon emission reduction, de facto nullifying any attempts in Europe and the United States.

In fact there is an associated danger that ironically lurks with the emerging situation. The public, unconvinced by emissions alarmism, may turn against all pollution control measures, throwing away the "baby with the bathwater." This could be a *bona fide* tragedy and there is evidence that this may be happening, for different reasons in both China and the United States. In China, public discourse about climate change and emission controls, without any substantive moves nearly close to what a Kyoto Protocol or successors would have liked, has already shifted the public debate from the real problems of air and water quality. People talk more about CO_2 rather than removing particulates from coal fired power plants which have totally obscured the skies of dozens of Chinese cities. In the United States, a Gallup poll in early April 2010 for the first time has shown that more than 50% of Americans want to develop US energy resources even if doing so will lead to environmental "suffering" of some kind.

However, changes and energy transitions will always happen. Such transformations will be painful and they must be rationalized by real economic motivations and the opportunity costs of not doing something.

Let's start with a forecast of world energy supply based on the EIA's forecast to 2030 and extrapolating to 2050, by using the EIA growth rates for each individual primary energy source. Table 12.1 (next page) provides the individual contributions in quads (quadrillion BTU), showing that over the next 40 years, world energy demand will increase by more than 80%. Liquids, dominated by oil will account for 258 quads, or about 117 million barrels per day, compared to about 85 million barrels per day in 2010. We think this is physically possible but clearly cumbersome and constantly burdened by geopolitical frictions and unsavory practices by several producing countries.

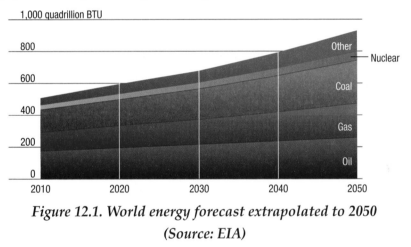

Figure 12.1. World energy forecast extrapolated to 2050
(Source: EIA)

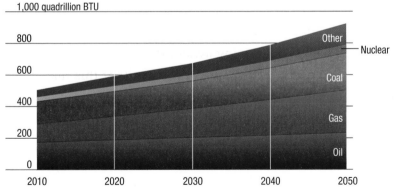

Figure 12.2. Adjusted world energy forecast with natural gas
replacing partial oil and coal growth (Source: EIA)

Table 12.1. Energy contributions from primary sources, forecast to 2050

Total World	2010	2020	2030	2040	2050
Liquids	174.7	194.2	215.7	**235.9**	**258.0**
Natural Gas	118.5	141.7	158.0	**185.2**	**217.0**
Coal	140.6	161.7	190.2	**225.1**	**266.5**
Nuclear	29.0	35.4	40.2	**47.1**	**55.2**
Other	45.6	62.8	74.1	**99.6**	**133.8**
Total	**508.3**	**595.7**	**678.3**	**792.9**	**930.6**

Figure 12.1 presents the same information in graphical form, showing oil to still be the premier fuel of the world economy (very near coal, each almost 30%) and that the three fossil fuels account for over 80% of the world energy demand by 2050.

We are both forecasting and proposing a desirable and plausible transformation. Natural gas, in abundant supply, not even counting yet the enormous potential of natural gas hydrates, can replace some oil and some coal in a movement that makes a lot of sense from a number of aspects:

- Natural gas has by the most recent estimates an ultimate recovery of over 30,000 trillion cubic feet, a 300-year supply at current uses
- Making inroads in the transportation sector can diversify its primary energy base. Gas can reduce oil dependency, a desirable event from a geopolitical point, and can readily reduce pollution. Because of existing developed infrastructure in the OECD countries, it is more logical to see a massive movement in the developing world, headed by China. As we have already mentioned there are more than 450 private cars per thousand people in the United States, while there are only 25 per thousand people in China. Natural gas in the transportation sector can be accomplished either directly through CNG vehicles or indirectly, through electrical cars. Indeed, as pointed out in Chapter 9, if serious efforts are made to adopt this approach using government subsidies of the sort used to promote alternative energy supplies, the impact will be significant and the benefits will begin to accumulate very rapidly. Replacing conventional liquid fuels with CNG can generate vast savings and contribute greatly to energy independence by reducing America's dependence on foreign oil. The greater the degree of conversion to CNG, at the expense of

gasoline and diesel, the greater is the impact. At the limit, a complete switch from gasoline and diesel to CNG can reduce America's daily oil consumption by 15 million BOPD. The $450 billion dollars thus saved per year (based on 2010 oil price), invested in the domestic economy rather than going to foreign governments, provide substantial funding for infrastructure upgrades and research. The reduction in fuel cost can also provide a major boost to the economy although, realistically, we expect natural gas prices to achieve parity with gasoline prices in the longer term.

• Replacing some coal from its expected growth in power generation is also a very desirable event, reducing pollution, especially particulates. Furthermore, natural gas is very attractive from an inverse economy of scale (smaller power units are attractive) compared to the ever increasing size of coal power plants. Much of the growth in power generation will have to be in developing and even remote areas of the world.

Thus, here is what we propose and forecast. Starting from 2015, 50% of the growth in non-OECD countries for liquids is to be replaced by natural gas. Also, 50% of the world would-be growth in coal is to be replaced by natural gas. Table 12.2 shows the expected energy amount to "change hands."

The resulting new world energy demand is shown in Table 12.3 and graphed in Figure 12.2. Nuclear and other forms of energy are left as forecast and extrapolated from the EIA figures. As can be seen in Figure 12.2, natural gas in this manner will emerge as the premier fuel of the world economy, accounting for 29% of the total by 2050 and eclipsing both oil and coal, reducing each to 25 %.

This transformation is possible, it is indicated and it is the most likely one to happen. It is also the one least disruptive, a major benefit in itself.

Table 12.2. Amounts in quads to be changed to natural gas in world energy demand (Source: World Energy Outlook)

Year	50% of non-OECD Oil Growth	50% of Total Coal Growth
2020	4.6	5.5
2030	11.3	17
2040	17.7	26
2050	23.4	33.6

Table 12.3. Adjusted primary world energy demand, forecast to 2050.
(Source: BP Statistical Review)

Total World	2010	2020	2030	2040	2050
Liquids	174.7	189.7	204.4	218.2	234.6
Natural Gas	118.5	151.8	186.3	228.8	274.1
Coal	140.6	156.2	173.2	199.2	232.8
Nuclear	29.0	35.4	40.2	47.1	55.2
Other	45.6	62.8	74.1	99.6	133.8
Total	508.4	595.8	678.2	792.9	930.6

Technologies of the Energy Future

It will certainly not be wind turbines or solar collectors that will dominate the real energy scene over the next several decades. Below we have selected two technologies that we think fit our vision of the future of energy. It is clear, and in contrast to the AEF report, that technology deployment in the energy business will continue to be dominated by fossil fuel applications in their extraction and with an additional emphasis on the diversity of their end uses.

Shale Gas

What has transformed shale gas, seemingly overnight, from a practically invisible resource to a mammoth natural gas reserve that can be tapped for centuries? The first commercial gas well, in 1821 in Fredonia, New York, was produced from a shale formation. And shale gas has been produced since the late nineteenth century in the Appalachian and Illinois Basins of the US. However, shale gas wells were typically marginal, and by the time of the Drake oil well discovery in 1859, shale gas waned in comparative significance to gas from conventional reservoirs. Not until the first decade of the twenty-first century, thanks to the Barnett Shale of east Texas, did shale gas really re-appear on the map.

In 1996, shale gas wells in the US produced 300 billion cubic feet of gas, still only 1.6% of total US gas production. By 2006, annual production was up to 1.1 trillion cubic feet, nearly 6% of US gas. In early 2010, the Barnett Shale alone produced 6% of US gas production, with US shales collectively accounting for 10% of total US natural gas. From 2000 to 2009, US gas shale production increased 8-fold, and together with coalbed methane, the combined contribution of these two sources is projected to account for 34% of US natural gas production by 2020, nearly doubling the current percent contribution of the two combined.[73]

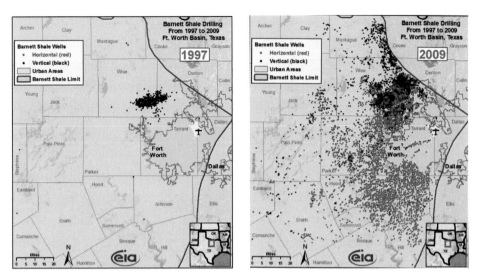

Figure 12.3. Barnett Shale Wells: 1997 and 2009 (Source: EIA)

The potential shale gas reserves picture is even more startling. Taking the case of the Marcellus alone, in the four years from 2005 to 2009, estimates of its gas reserves have increased from about 1 trillion cubic feet to over 500 trillion cubic feet, on the high end. There has never been before an upgrade of such magnitude in any underground hydrocarbon resource in such a short time. Add a possible 300 trillion cubic feet from the more economically challenged Haynesville Shale. Plus, there exist the next gas shale frontiers anticipated throughout Europe, China, and Australia, among others. The Russian business daily Kommersant has reported that Europe's gas reserves could jump 47% from early 2010 levels if Poland's gas reserves (~100 trillion cubic feet) are confirmed and drilling for shale gas is already underway in several other countries in mainland Europe. In 2009, Russia supplied around 25% of Europe's gas but during both 2006 and 2009 it withheld supplies to certain countries over pricing disputes. This highlighted Europe's vulnerability to a re-emergent Russia so if these shale prospects come to fruition they could have major financial and political implications. Poland alone could satisfy its own domestic demand beyond the next 200 years, an extremely significant development considering it imports 70% of its current gas requirement from Russia. The potential figures, in North America and abroad, are astounding.

So, what is behind the overnight sensation of shale gas – its transformation from small time also-ran to emerging leader of the clean, green, natural gas world of the future? It is simply the successful deployment of technology. And it is not with absolutely new, until recently, unimaginable technology, but with a combination of existing, albeit improving, technologies and methods that we discussed initially in Chapter 4. These include improved well logging methods to identify the most productive shale zones, and advancements in directional drilling of long horizontal wells with lateral lengths to 10,000 feet and greater. Gone are the days of low producing vertical shale gas wells. Advancements in hydraulic fracturing technologies are especially at play, enabling multiple and long created fractures along the horizontal length of the well, and extending through the otherwise unproductive or poorly productive shale. Of increasing use are custom horizontal well injection "tool strings" that can be run into the well so that more fractures can be induced in less time, and spaced apart as desired, to "connect" with more of the gas reservoir. More efficient, high horsepower pumps that can inject at high rate, simplified yet more effective fracturing fluids, utilizing special proppants and chemical additives designed to enhance and maintain high initial gas production rates from new wells. Two or more wells can be fracture stimulated at the same time. Multiple horizontal wells can be drilled from a single drilling pad. More fractures – more gas – lower cost.

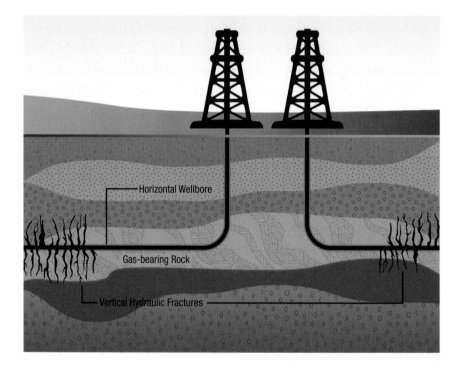

Figure 12.4. Depiction of simultaneous fracturing of two horizontal gas shale wells (Source: Regal Energy)

The shale gas success story is a spectacular, but realistic example of what purposeful and effective deployment of technology can accomplish. The implications for the future are profound.

The Next Generation Gasifier

One of the most important shortcomings of energy resources is their lack of overlap for end users. We have expounded upon this theme in this book. In 2010 virtually no oil was used in power generation and virtually nothing but oil for transportation. Energy diversity becomes an important component of the future energy mix. The ability to convert solid hydrocarbons such as coal, pet coke and even biomass into synthesis gas, a mixture of carbon monoxide and hydrogen, offers a number of options such as direct use in electricity generation of a more flexible fuel, the production of hydrogen for fuel cell cars and more importantly, the use of synthesis gas (syngas) to produce a series of liquid products from alcohols (primarily methanol) to jet fuel and, very clean, diesel.

Table 12.4. New Gasifier: Cost Advantages for Electric Power Generation

	ZEEP	Cost Increase over ZEEP	
		General Electric	Shell Oil
Total Plant Cost, $/kW	1211	+16.8%	+27.7%
Cost of electricity, $/kW-hr	0.042	+13.5%	+21.0%
Cost of hydrogen, $/Mcf (million cubic feet)	2.547	+25.2%	NA

The technology of converting hydrocarbons into syngas and then to liquids, often referred to as gas-to-liquids (GTL), coal-to-liquids (CTL) or generally XTL is not new and has its origin in Germany between the two world wars (Fischer-Tropsch method) and expanded in South Africa during the apartheid era. Around 2000 Boeing embarked upon a mission to expand the use of its leading edge advanced aerospace technologies into the energy business. They decided that their most unique technological edge was in gasification based on Boeing's Rocketdyne Division's expertise in solid rocket fuel technology. This led to a compact, low-cost gasifier for not only coal, but also petroleum coke ("petcoke") and biomass as fuel.

In 2009, now in a joint venture, the company successfully completed pilot-scale testing of a 400 tons-per-day feed system that will support the design of a 3,000-tons-per-day coal equivalent commercial scale plant with clear advantages over other gas¬ifi-cation technologies. Figure 12.5 shows clearly the fundamental advantage of the new gasifier compared to the two other major gasifiers deployed by the industry.

A study contacted by an independent engineering firm found that the new technology has significant economic advantages over the two established technologies, shown in Table 12.4. These differences when applied to commercial size plants would amount to at least $1 billion.

Important additional advantages include greater reliability and availability. Project engineers from the US Department of Energy ("DOE") estimate 99% availability due to the long life components used in construction, rapid repair capabilities, smaller size and lesser complexity, and significantly shorter scheduled outages. Typically, conventional gasifiers require backup units to achieve availability greater than 85%. Another important factor is feedstock tolerance. Due to its unique "Dry Pump" feed system, the Compact Gasifier is expected to allow much greater flexibility in feedstocks. Other gasifiers are slurry-feed which can create difficulties with biomass, lignite, and even sub bituminous coals, the largest source of coal in the US. Finally, when it comes to gasifiers it appears that size is indeed important. Due to the Compact Gasifier's ultra-high heating rates, temperatures and solid fuel handling, the size of the gasifier is only about 1/10th that of conventional gasifiers.

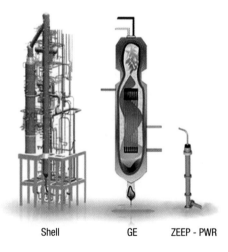

Shell   GE   ZEEP - PWR

Figure 12.5 Rocketdyne-ZEEP Gasifier Compared to Current Technologies

This is only one more example of the type of breakthrough that can lead to the wide scale adoption of an already familiar technology. Yet, the potential significance of using an efficient, compact device to transform all manner of raw materials into syngas and thence to liquid fuels or feedstock for other industries should not be underestimated. Potentially, the gasifier could convert corn hulls and straw into syngas, leaving the crop itself to be used as food and producing liquid fuels with already-proven technology.

Yesterday's Addiction – Tomorrow's Necessity

From the first discovery of natural gas in ancient Persia, to the first use of the "stone that burns" in China to smelt copper some 3,000 years ago, to the bamboo-drilled oil-bearing wells in fifth century China, to the industrial revolution fueled by coal, to the first commercial gas wells and spectacular oil strikes of the nineteenth century, to the twentieth century discoveries of giant oil and gas fields, to the mind-boggling "extreme" well constructions of today, both onshore and offshore – all has been enabled by remarkable creativity and advancements in "oilfield technology." As a result, hydrocarbons continue to be abundant and affordable and the biggest energy source for mankind, despite the false odds and incorrect bets placed against them.

Throughout this book, we have tried to balance arguments on all sides of the issues that we have discussed. We believe that the world is indeed in the process of diversifying its energy supply and that the change will be realized over many decades, like most other major technological or societal changes that have happened in the past. In Chapter 8, we discussed the electric power industry and how it blossomed and grew, despite years of debates on standards and which system was technically superior. Once the implementation began, however, it was only a few years until electricity had spread out to change the vocabulary of power and convenience, forever. Yet, 120 years on, there are huge segments of the global population who still have no access to electricity.

So it will be for energy change. In some locations, electric cars will carry riders though the streets in a few years time, while in others, vehicles powered with natural gas (CNG), blends of biodiesel, GTL, or ethanol or just plain old gasoline will putter along for years to come. Some countries will take advantage of HVDC power grids to export surplus power to others while some of their rural inhabitants will still rely on a water wheel or an old diesel generator to provide essential utilities to which they may have only recently gained access.

But that, in fact, is the strength and robustness of this transition process. There is no "one size fits all" descriptor for energy use. So, logically there should not be only one single solution. The world has begun to move inexorably towards renewable, or more sustainable and, potentially, more diverse sources of fuel and energy. That's not a bad thing since we will still rely on hydrocarbons for the bulk of our energy supply for a long time to come. If renewables can provide even a small, but steadily increasing, fraction of our energy requirements, at reasonable cost, they will ultimately help extend the longevity of our hydrocarbon reserves. Despite the rantings of environmentalists, politicians and the media, we should not forget the enormous debt we owe to oil, gas and coal for providing the bulk of the energy required to lift us from the subsistence agrarian economy of 1700 to the sophisticated global society that we have become. Nor should we squander everything we have achieved to date, tilting at windmills (no pun intended) of the type represented by AGW and, in the process, putting the health and wealth of billions of the planet's inhabitants, present and future, at stake.

We have important work ahead of us – lifting two billion people out of poverty, providing electricity, clean water and sanitation, investing in education and health care, helping establish agricultural and industrial infrastructure, and fostering the political and financial stability needed for the disadvantaged to grab a stake in the world economy. To constrain growth by imposing spurious taxes, siphoning-off money from the real economy, dampening business activity and entrepreneurial talent, hardly seems like a good formula for success, unless you happen to be one of the potential beneficiaries of that cash – politicians, bankers, taxmen, climate researchers, banana republic despots and any number of shrewd operators who will figure out how to game the system, as is already being done on the carbon markets.

We should not be distracted by prophets of doom, by misguided do-gooders or by those seeking to line their pockets with free cash from taxpayers. Instead, we should use our still abundant and affordable hydrocarbon resources to help us power the world, generating, and better distributing, the wealth needed to tackle problems along the way and accomplish our long term goals. Hydrocarbons sparked the old world and fueled the modern world. And, led by clean and green natural gas, hydrocarbons are poised to save the future of the world. If we can just get past the rhetoric, that is the reality.

That is, in fact, The Energy Imperative.

APPENDICES

Appendix A
A Few Energy Facts

General

– The developed world (OECD) uses one-half of the world's total energy to produce one-half of the world's GDP

– 80% of world population is projected to be in non-OECD countries by 2030

– In China there is one car for every 25 people

– In the United States there is about one car for every 2 people

– The largest non-fossil fuel source is biomass burned for heat (mostly wood and dung).

– It is estimated that the US could produce up to 4 billion BOE from biomass

– Current total fossil fuel reserves: over 14 trillion barrels of oil; 66 quadrillion cubic feet of natural gas; 900 billion tons of coal

– From the smallest to the largest in size, there are over 40,000 oil and gas fields in the world

Natural Gas

– Qatar is the biggest producer and exporter of Liquified Natural Gas (LNG) – in 2008 it exported over 1.4 trillion cubic feet – about 80% of which went to South Korea, Japan, and India

– Iran and Qatar combined own 73% of the natural gas reserves in the Middle East

– The US is the world's biggest natural gas producer – nearly 21 trillion cubic feet in 2009

Oil

- Nearly 95% of oil is produced from only the 1,500 largest oilfields

- World oil production is ~85 million barrels per day (~3.57 billion gallons per day) or about 1,000 barrels per second

- The total volumetric consumption of oil is a little over 1 cubic mile per year

- The world consumes 200 million gallons of jet fuel per day

- US diesel fuel consumption is around 60 billion gallons per year

- 18 billion barrels of oil are shipped each year by tanker

- It can take 15-20 years to go from exploration to production in new, large oil discoveries

- Production from a new, large oil discovery can last over 50 years

- The first offshore oil platform was constructed in 1947 by the Superior Oil Company off the coast off the Gulf Coast of Louisiana

Coal

- Coal is second to oil as the greatest contributor to the worldwide energy mix

- Worldwide coal resources are greater than oil resources – primarily from the US and China

- Coal generates about 50% of US electricity supply and 80% of China's

- Annual worldwide coal consumption was 6.7 billion tons in 2008

- China accounted for 47% of coal consumed globally in 2009

Electricity

- The US has the biggest installed capacity for wind power (35 GW in 2009) up from only 9 gigawatts in 2005 but this generates less than 2% of the country's electricity.

- Denmark generates 20% of its electricity from wind power.

- The US is the biggest producer of geothermal power and has 3 gigawatts of installed capacity

- World electricity consumption is around 2 terawatts

- Iceland has the highest electricity consumption per capita – over 30,000 kilowatt-hours – about 2.5 times greater than the figure for the US

- In China, electricity consumption is only ~2,500 kilowatt-hours per capita – about 5 times less than the average for the US

- Recently reported data (2006) indicated 1.6 trillion kilowatt-hours of electricity were lost in transmission worldwide – representing 9% of electricity generated

CO_2 Emissions

- On a per capita basis, Australia has the highest CO_2 emissions per year (25.3 tons CO_2 per person) – 5 times greater than China

- On a country basis, China has the highest CO_2 emissions per year – greater than 8.25 billion tons – almost 20 times greater than Australia and 5 times greater than Russia

Nuclear

- Nuclear energy contribution to the total worldwide energy mix is 6% (16% of electricity)

- Nuclear power can come from the fission of uranium, plutonium or thorium or the fusion of hydrogen into helium. Today it is almost all uranium.

– One fuel pellet the size of a pencil's eraser produces about the same amount of energy as burning 1 ton of coal, 150 gallons of oil or 17,000 cubic feet of natural gas.

– Nuclear power supplies France with nearly 80 percent of the country's electricity.

– Over 160 million people in the US live within 75 miles of the Department of Energy's nuclear waste storage sites

– The average amount of radiation absorbed by US residents annually is 360 millirems. The total extra radiation that neighbors of Yucca Mountain would absorb would be 20 millirems over the 24-year projected life of the power plant.

– Over a ten year period beginning in 2015, it is projected that one new nuclear reactor will start up every 5 days in Asia.

Alternatives – Renewables

– As recently as ~150 years ago, wood, one form of biomass, supplied up to 90% of energy needs

– Wind and solar energy provide only 1% of the total worldwide energy mix

– Hydroelectric power contribution to the total worldwide energy mix is only 2%

– Although growing rapidly, liquids from biomass (for example ethanol from corn and sugar cane) contribute only 1% of the energy provided by oil

– The US is the biggest producer of ethanol – about 10.75 billion gallons in 2009

– Hydrogen has only 29% of the energy of gasoline on a volume for volume basis

Appendix B
Physical Classification of Crude Oil

As mentioned in Chapter 1, the physical classification of crude oil is generally based on its API (American Petroleum Institute) gravity. Oil gravity is a measurement of its density relative to water.

The formula for calculating API gravity is:

API gravity (degrees API) = (141.5 / SG) – 131.5
where SG is the specific gravity of oil at 60 ºF

Specific gravity of water, the relative standard, is 1. So, oil with the same density as water has an API gravity of 10.0 degrees API. The API classifies oil as follows:

Light oil	*– API gravity greater than 31.1 ºAPI*
Medium oil	*– API gravity between 22.3 ºAPI and 31.1 ºAPI*
Heavy oil	*– API gravity below 22.3 ºAPI*
Extra heavy oil	*– API gravity less than 10 ºAPI*

In crude oil markets, prices are based on a limited number of benchmark crude oils. All other crude oils are then priced at a discount or premium to these benchmarks.

The primary benchmark crude oils are:

Brent
Blend of crudes produced from North Sea fields. Brent is not nearly the most significant benchmark in terms of sales volume, but it is used to set pricing of about two-thirds of produced crudes. Brent is light (high API gravity) and sweet (low sulfur content).

West Texas Intermediate (WTI)
WTI is the United States benchmark. It is of high quality, light and sweet. It is lighter than Brent, and thus commands a premium in its price.

OPEC Basket

The OPEC Basket is an average of about 15 different crudes from OPEC (Organization of Petroleum Exporting Countries) nations. Its average quality (gravity and sulfur content) is lower than Brent and WTI. This benchmark enables OPEC to have some control in keeping pricing within a certain range, over given periods.

Dubai

Dubai (or Dubai and Oman) – also known as Fateh – is a benchmark to set pricing of Dubai crude oil exports to Asia. This benchmark was also created because Dubai crude is one of the few from the Persian Gulf that is available in spot markets (rather than requiring long term supply contracts).

Appendix C
Chemical Classification of Hydrocarbons in Crude Oil

Crude oil hydrocarbons are almost entirely comprised of the following four hydrocarbon types:

- Paraffins *(alkanes)*
- Olefins *(alkenes)*
- Naphthenes *(cycloalkanes)*
- Aromatics *(benzenes)*

The relative amounts of each hydrocarbon type, and individual molecular sizes within each type, give each crude oil its unique set of physical and chemical properties. Additionally, there are the non-hydrocarbon impurities, incorporated in the non-paraffin components, to varying degrees. Their presence is especially important in the refining processes.

Paraffins
A basic paraffin (or alkane) structure – for octane – is as follows:

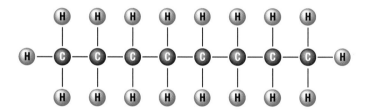

Figure C.1. Basic example paraffin – octane (C_8H_{18})
(Black spheres are carbon, blue spheres are hydrogen)

The paraffinic series of hydrocarbon compounds (also known as alkanes) found in crude oil have the general formula C_nH_{2n+2}. They can be either straight chains (normal) or branched chains (isomers) of carbon. They range from C_5 to C_{50+}. In paraffins the carbon atoms are joined only by single C-C bonds (saturated).

Olefins

Olefins (alkenes) are also present in certain crude oils. They are unsaturated hydrocarbons in which at least one pair of carbon atoms is joined by a double bond (C=C). This higher degree of carbonization gives olefins one of their defining characteristics – they burn with a bright smoky flame. Olefins have been identified in many crude oils and may reach concentrations of 7% to 8% in crudes from certain geographical locations (e.g. many Russian crude oils). Following is a simple olefin (pentene) structure:

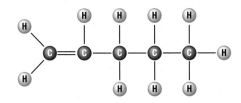

Figure C.2. Basic example olefin (alkene) – pentene (C_5H_{10})

Naphthenes

Naphthenes (also known as cycloalkanes) are saturated (single C-C bonded) hydrocarbon structures in the form of closed rings.

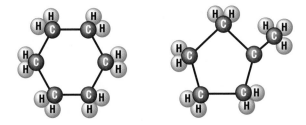

Figure C.3. Example naphthene (cycloalkane) structures

Naphthenes have the general formula C_nH_{2n}. The carbon chain, or backbone, forms cyclic rings with each carbon atom surrounded by the associated hydrogen atoms. Naphthenes may be a single ring with five or six carbon atoms. Two-ring structures (dicycloparaffins) are found in heavier crudes.

Aromatics

Aromatics (benzenes) are also ring compounds – but unsaturated.

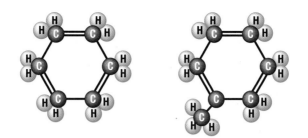

Figure C.4. Example basic aromatic (C6 ring) compound structures

Basic aromatics can have pleasant aroma, thus the name. However, that association does not generally apply to crude oils.

Heavier crudes contain multi-ring (complex) aromatics, known as polynuclear aromatics. These contain three or more "fused" aromatic rings. These structures, such as asphaltenes, especially characterize heavy crudes. While asphaltene structures are not fully known, an example of a portion of a typical asphaltene is shown in Figure C.5.

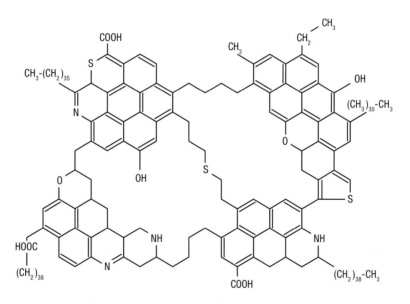

Figure C.5. Portion of a typical asphaltene

Such complex polynuclear hydrocarbons are of high molecular weight, thus increasing the density (and lowering the API gravity) of the crude oil – and increasing its viscosity. They also contain impurities such as sulfur, nitrogen, oxygen, and heavy metals. The combination of the large molecular size (and high C:H ratio) and the presence of impurities increases the complexity and cost of refining crudes containing higher levels of polyaromatics.

Appendix D
Oil Refinery Units and Processes

This appendix supplements Chapter 6, to provide basic information on oil refinery units and processes.

Desalter Unit

Before crude oil enters the Crude Distillation Unit (DCU), it is preheated using heat from refinery product streams, the salt is washed out, and the crude is dehydrated using electrostatic enhanced liquid/liquid separation.

Atmospheric Distillation Unit and Vacuum Distillation Unit

There are two steps in the distillation process – first through the Atmospheric Distillation Unit and then through the Vacuum Distillation Unit. In atmospheric distillation, crude oil is heated to the desired temperature using fired heaters and is then flashed to its fractions in the distillation column. The lighter, volatile (low boiling point) components come off the top of the column, while the heavy (high boiling point) components come off the bottom. The various fractions are then distributed to the different conversion and treating units.

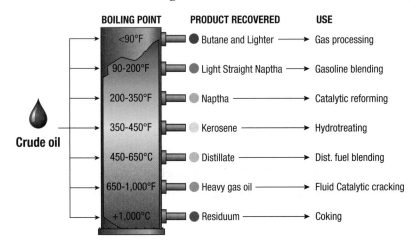

Figure D.1. Crude Distillation Unit Product Ends (Source: EIA)

Typically, the light ends (butane and lighter) are about 2-3% of the yield of the CDU. Light naphtha is about 6%, naphtha (medium and heavy naphtha) total about 25%, kerosene about 10%, heavy gas oil about 15%, and residuum over 40% or more.

In vacuum distillation, the residuum from the CDU is heated to a desired temperature, and flashed in the VDU column. The end result is a collection of gas oils that are then further refined (converted) to primarily fuel products. Remaining heavier ends go to coking units – to make additional gasoline, diesel, and solid carbon coke.

Catalytic Reformer Unit

The catalytic reformer unit contains catalyst that enables conversion of the naphtha-boiling range molecules into higher octane reformate (reformer product). Reformate has higher content of aromatics, olefins, and cyclic hydrocarbons. An important byproduct of a reformer is hydrogen.

Fluid Catalytic Cracking (FCC) Unit

The FCC upgrades heavier fractions into lighter, more valuable products, such as naphtha (precursor to gasoline) and diesel. FCC units utilize a regenerated / recycled solid catalyst on which thermal cracking of the feedstock takes place. Catalysts are very specifically modified zeolites. Zeolites are alumino-silicates that have cage-like porous structures – giving them special applications. There are many naturally-occurring zeolites, and even more synthetic zeolites. Synthetic zeolites are used as molecular sieves (filtration) and as oil refinery catalysis. Zeolites have refining catalysis applications because they can be shape-selective. The zeolite catalysts, with their particular cage opening size and pore space size, can selectively allow or exclude reactants and reaction products based on molecular diameter – to result in a specific mix of end products. This size and shape selectivity is pre-designed in a synthetic zeolite catalyst. Also, because of their extremely high surface area, zeolites can be used as the supporting (impregnated) medium for the active metals that catalyze refining reactions.

Gasoline produced from the FCC has a high octane rating but contains olefins (unsaturated hydrocarbons having C=C double bonds). The presence of olefins in gasoline leads to deposits during combustion. Therefore, the gasoline fraction is subject to further refining, to "saturate" (hydrogenate) the olefins.

Hydrocracker Unit

The hydrocracker uses hydrogen to upgrade heavier hydrocarbon fractions (and unsaturated fractions) into lighter, more valuable products. The catalysts are silica-alumina or zeolites impregnated with certain metal combinations (such as nickel-tungsten and nickel-molybdenum). The major output products from the hydrocracker are low sulfur jet fuel and diesel. Gasoline and LPG are also produced. Hydrocracking is a preferred process in India, Asia, and Europe, given their higher demands for diesel (and kerosene). The FCC unit has greater utility in the US, as it is the workhorse for the eventual production of finished gasolines.

Figure D.2. Massive Repsol hydrocracker in Spain (Source: Siemens)

Coking Units

The coking units (cokers) process heavy residuum (or resid) into gasoline and diesel fuel, leaving coke as a residual product. About 20-35% of the yield is coke. Coke is solid and close to pure carbon. The heavy residual is often viewed as the undesirable portion from distillation – as a necessary evil. However, the coking process leads to valuable specialized (non-fuel) products. Coke is used to manufacture carbon anodes and highly crystalline needle coke (graphite). It is also used in the production of carbon steel alloys and as an additive to rubber and plastic. Otherwise, coke is used for power generation.

Figure D.3. HOVENSA Coker Unit – St. Croix (Source: Hess Corporation)

Figure D.4. Petroleum coke

Isomerization Unit

This unit converts linear hydrocarbons into branched hydrocarbons that burn more efficiently – having a higher octane number. These branched hydrocarbons (isomers) can be blended into gasoline fractions. Octane number of gasoline is an indication of its knock resistance. Compressed gasoline-air mixtures have a tendency to ignite prematurely in internal combustion engines – rather than burning smoothly. This causes engine "knock," which is characterized by a pinging sound in one or more cylinders. The octane number of an actual gasoline is based on comparison to isooctane (2,2,4-trimethylpentane) and heptane. Isooctane is a branched hydrocarbon that burns smoothly with minimal knocking. Thus, it is arbitrarily assigned an octane number of 100. Heptane, which is a straight chain (unbranched) hydrocarbon, burns inefficiently and is "high-knocking." Therefore, it is assigned an octane rating of zero. If a particular gasoline has an octane rating of 90 that means its knocking characteristic would be the same as a 90% isooctane – 10% heptane mixture.

Appendix E
Gas Processing

As mentioned in Chapter 6, gas processing involves removal or separation of impurities from natural gas (methane, CH_4). There are four basic processes:

- Oil and gas condensate removal
- Water removal
- Separation of Natural Gas Liquids (NGLs)
- Hydrogen sulfide (H_2S) and carbon dioxide (CO_2) removal

H_2S and CO_2 removal are covered in Chapter 6. The first three processes are detailed further, below.

Oil and Condensate Removal

This stage is typically conducted at the wellhead with "separators" that rely on gravity to separate heavier oil and condensate from the lighter gas. If this separation stage is not enough, then specialized, low temperature separators are used. These rely on pressure reductions which allow the gas to expand, resulting in the familiar Joule-Thomson cooling that is the basis of modern refrigeration. This cooling of the condensate and oil-laden natural gas causes "drop out" of the hydrocarbon liquids from the purely gaseous components.

Water Removal

Liquid water is also mostly removed with simple separation methods at or near the wellhead. Removal of water vapor – dehydration – is more complex. The two dehydration process options are absorption and adsorption. The absorption method uses a liquid dehydrating agent (or desiccant) – glycol. The adsorption method condenses water vapor on a solid desiccant – activated alumina or granular silica. Water adsorbs onto the solid surfaces, from which it is then collected.

Separation of Natural Gas Liquids (NGL)

Produced natural gas contains natural gas liquids (NGLs). NGL is typically more valuable as stand-alone product. Its removal is desirable – both for the sake of dry methane and because it is economically justified for the separate NGL product. The two primary methods for removing NGLs from raw natural gas are the absorption method and the cryogenic expander process. The absorption method for NGL extraction is very similar to the method for dehydration – except that oil that has an affinity for NGLs is used in place of glycol which has an affinity for water. Cryogenic processes are used to extract the lightest non-methane hydrocarbons (such as ethane) from the natural gas. Cryogenic processes rapidly drop the temperature of the gas stream – to about -120 ℉ (-84 ℃) – at which point ethane and other light-end hydrocarbons condense and drop out while methane remains in gaseous phase.

Figure E.1. Absorption Towers (Source: Duke Energy Gas Transmission Canada)

Appendix F

Nuclear Power Plants in Commercial Operation and Planned (as of June 2010)

Country	Nuclear Electricity Generation 2009		Reactors Operable		Reactors Under Construction		Reactors Planned		Reactors Proposed		Uranium Required 2010
	Billion Kwh	% E	No.	Mwe	No.	Mwe	No.	Mwe	No.	Mwe	Metric Tons U
Argentina	7.6	7.0	2	935	1	692	2	767	1	740	123
Armenia	2.3	45	1	376	0	0	1	1060			55
Bangladesh	0	0	0	0	0	0	0	0	2	2000	0
Belarus	0	0	0	0	0	0	2	2000	2	2000	0
Belgium	45	51.7	7	5943	0	0	0	0	0	0	1052
Brazil	12.2	3.0	2	1901	0	0	1	1245	4	4000	311
Bulgaria	14.2	35.9	2	1906	0	0	2	1900	0	0	272
Canada	85.3	14.8	18	12679	2	1500	4	4400	3	3800	1675
China	65.7	1.9	11	8587	23	25310	34	38160	120	120000	2875
Czech Republic	25.7	33.8	6	3686	0	0	2	2400	1	1200	678
Egypt	0	0	0	0	0	0	1	1000	1	1000	0
Finland	22.6	32.9	4	2696	1	1600	0	0	1	1000	1149
France	391.7	75.2	58	63236	1	1630	1	1630	1	1630	10153
Germany	127.7	26.1	17	20339	0	0	0	0	0	0	3453
Hungary	14.3	43	4	1880	0	0	0	0	2	2200	295
India	14.8	2.2	19	4183	4	2572	20	16740	40	49000	908
Indonesia	0	0	0	0	0	0	2	2000	4	4000	0
Iran	0	0	0	0	1	915	2	1900	1	300	148
Israel	0	0	0	0	0	0	0	0	1	1200	0

Continued on next page

Country	Nuclear Electricity Generation 2009		Reactors Operable		Reactors Under Construction		Reactors Planned		Reactors Proposed		Uranium Required 2010
	Billion Kwh	% E	No.	Mwe	No.	Mwe	No.	Mwe	No.	Mwe	Metric Tons U
Italy	0	0	0	0	0	0	0	0	10	17000	0
Japan	263.1	28.9	55	47348	2	2756	12	16532	1	1300	8003
Jordan	0	0	0	0	0	0	1	1000	2	600	0
Kazakhstan	0	0	0	0	0	0	2	600	2	600	0
Korea Dpr (*North*)	0	0	0	0	0	0	0	0	1	950	0
Korea Ro (*South*)	141.1	34.8	20	17716	6	6700	6	8190	0	0	3804
Lithuania	10.0	76.2	0	0	0	0	0	0	2	3400	0
Malaysia	0	0	0	0	0	0	0	0	1	1200	0
Mexico	10.1	4.8	2	1310	0	0	0	0	2	2000	253
Netherlands	4.0	3.7	1	485	0	0	0	0	1	1000	107
Pakistan	2.6	2.7	2	400	1	300	2	600	2	2000	68
Poland	0	0	0	0	0	0	6	6000	0	0	0
Romania	10.8	20.6	2	1310	0	0	2	1310	1	655	175
Russia	152.8	17.8	32	23084	10	8960	14	16000	30	28000	4135
Slovakia	13.1	53.5	4	1760	2	840	0	0	1	1200	269
Slovenia	5.5	37.9	1	696	0	0	0	0	1	1000	145
South Africa	11.6	4.8	2	1842	0	0	3	3565	24	4000	321
Spain	50.6	17.5	8	7448	0	0	0	0	0	0	1458
Sweden	50.0	34.7	10	9399	0	0	0	0	0	0	1537
Switzerland	26.3	39.5	5	3252	0	0	0	0	3	4000	557
Thailand	0	0	0	0	0	0	2	2000	4	4000	0

Continued on next page

Country	Nuclear Electricity Generation 2009		Reactors Operable		Reactors Under Construction		Reactors Planned		Reactors Proposed		Uranium Required 2010
	Billion Kwh	% E	No.	Mwe	No.	Mwe	No.	Mwe	No.	Mwe	Metric Tons U
Spain	50.6	17.5	8	7448	0	0	0	0	0	0	1458
Sweden	50.0	34.7	10	9399	0	0	0	0	0	0	1537
Switzerland	26.3	39.5	5	3252	0	0	0	0	3	4000	557
Thailand	0	0	0	0	0	0	2	2000	4	4000	0
Turkey	0	0	0	0	0	0	4	4800	4	5600	0
Ukraine	77.9	48.6	15	13168	0	0	2	1900	20	27000	2031
UAE	0	0	0	0	0	0	4	5600	10	14400	0
UK	62.9	17.9	19	11035	0	0	4	6600	6	8600	2235
USA	798.7	20.2	104	101163	1	1180	9	11800	23	33000	19538
Vietnam	0	0	0	0	0	0	4	4000	6	6000	0
World**	2560	14	439	374,690	57	57,555	151	165,699	345	366,775	68,646
	Billion Kwh	% E	No.	Mwe	No.	Mwe	No.	Mwe	No.	Mwe	Metric Tons U

Sources: Reactor data: WNA to 3/6/10

IAEA- for nuclear electricity production & percentage of electricity (% e) 3/5/10.

WNA: Global Nuclear Fuel Market (reference scenario) - for U.

Operating = Connected to the grid

Building/Construction = first concrete for reactor poured, or major refurbishment under way

Planned = Approvals, funding or major commitment in place, mostly expected in operation within 8-10 years

REFERENCES

1. *BP Statistical Review of World Energy 2006*
2. *Annual Energy Outlook 2010,* US Energy Information Administration (EIA); https://www.sustainablebusiness.com/index.cfm/go/news.display/id/20303
3. http://www.deepwater.com/fw/main/Transocean-GSF-Rig-127-Drills-Deepest-Extended-Reach-Well-283C4.htm
4. *BP Statistical Review of World Energy 2010*
5. Modern Shale Gas Development in the US: A Primer (US Department of Energy, 2009)
6. World Bank Data Bank (data.worldbank.org)
7. Oil in the Sea III: Inputs, Fates, and Effects (http://books.nap.edu/catalog/10388.html)
8. Central Intelligence Agency (2007). *CIA World Factbook 2008* (Skyhorse Publishing)
9. Huber, Mark (2001). *Tanker operations: a handbook for the person-in-charge (PIC).* Cambridge, MD: Cornell Maritime Press
10. Bureau Veritas and Gas Carriers (September 2007) (http://www.newsletterscience.com/marex/pdf/00000153.pdf)
11. Mitsui Shipbuilding Nears Practical Application of Natural Gas Hydrate Transport Technology (http://www.japanfs.org/en/pages/026000.html)
12. http://web.worldbank.org/WBSITE/EXTERNAL/TOPICS/EXTOGMC/EXTGGFR/0,,menuPK:578075~pagePK:64168427~piPK:64168435~theSitePK:578069,00.html
13. US Energy Information Administration (www.eia.doe.gov)
14. US Energy Information Administration (www.eia.doe.gov)
15. Citgo (www.citgo.com); US Energy Information Administration (www.eia.doe.gov)
16. US Energy Information Administration (www.eia.doe.gov)
17. *Coal Information 2009,* International Energy Administration (www.iea.org); World Coal Institute (www.worldcoal.org)
18. US Department of Energy (www.energy.gov)
19. US – Canada Power System Outage Task Force – August 14th Blackout: Causes and Recommendations (http://www.nerc.com/docs/docs/blackout/ch5.pdf)

20. Swiss Federal Office of Energy: Report on the blackout in Italy on 28 September 2003 (November 2003)

21. Electricity Storage Association (http://www.electricitystorage.org/site/technologies/pumped_hydro/)

22. World Nuclear Association – World Nuclear Power Reactors & Uranium Requirements (http://www.world-nuclear.org/info/reactors.html)

23. Geothermal Energy: International Market Update (Geothermal Energy Association, May 2010)

24. High Voltage Direct Current (HVDC) Transmission Systems Technology Review Paper (http://www.twolf.com/pub/energy/technology_abb.pdf)

25. US Energy Information Administration (http://www.eia.doe.gov/basics/quickoil.html)

26. http://www.after-oil.co.uk/runways.htm

27. US Department of Energy (http://www.afdc.energy.gov/afdc/fuels/emerging_biobutanol.html)

28. BP – DuPont Biofuels Factsheet (http://www.bp.com/liveassets/bp_internet/ globalbp/STAGING/global_assets/downloads/B/Bio_bp_dupont_fact_sheet_ jun06.pdf)

29. *Nature* 447, 982-985 (21 June 2007), "Production of dimethylfuran for liquid fuels from biomass-derived carbohydrates", Yuriy Román-Leshkov, Christopher J. Barrett, Zhen Y. Liu, and James A. Dumesic

30. http://www.biodiesel.org/resources/reportsdatabase/reports/gen/ 20011101_gen-346.pdf

31. http://www.biodiesel.org/resources/fuelfactsheets/default.shtm

32. Neste Oil (http://www.nesteoil.com/default.asp?path=1,41,535,547,3716,3884)

33. National Renewable Energy Laboratory: "The Potential for Biofuel from Algae", Philip T. Pienkos (Algae Biomass Summit, San Francisco, CA, November 15, 2007). http://www.nrel.gov/docs/fy08osti/42414.pdf

34. http://cleantech.com/news/2076/tyson-syntroleum-to-build-biodiesel-plant

35. http://www.changingworldtech.com/

36. Green Car Congress: MIT Researchers Develop Lithium Iron Phosphate Material with Charge/Discharge Rates Comparable to Supercapacitors(March 12, 2009) (http://www.greencarcongress.com/2009/03/mit-researchers.html)

37. The World Bank – World Development Indicators (WDI 2010) – Population (http://data.worldbank.org/data-catalog/world-development-indicators/ wdi-2010)

38. *Silent Spring*, Carson, Rachel (Boston: Houghton Mifflin, 1962), Mariner Books, 2002, ISBN 0-618-24906-0

39. http://www.junkscience.com/malaria_clock.html

40. http://www.incidentnews.gov/incident/6250 Ixtoc-1 blowout

41. *Nature* 290, 235 - 238 (19 March 1981) (http://www.nature.com/nature/journal/v290/n5803/abs/290235a0.html)

42. *DEMOLITION AND SABOTAGE OF KUWAIT'S OIL INFRASTRUCTURE*, Edward V. Badolato and Robert F. Dexter (June 1992) (http://www.iiainc.net/file.php?cmd=view&id=92)

43. G. Landrey, Wilbur: "Oil slick in gulf likely to spread", *St. Petersburg Times* (30 January 1991), p. 5A

44. http://www.oilspillresponse.com/pdf/Educational/amoco.pdf

45. Bob Grant: "Oil spill is boon to bacteria", TheScientist.com (25 May 2010) (http://www.the-scientist.com/blog/display/57448/)

46. http://www.braer.net/

47. *The Skeptical Environmentalist : Measuring The Real State Of The World*, Bjorn Lomborg (Cambridge University Press 2003) ISBN 0-521-01068-5 (pp 192)

48. http://wiki.answers.com/Q/How_many_chickens_are_killed_each_day_for_consumption_in_the_us

49. http://en.wikipedia.org/wiki/Lakeview_Gusher

50. Oil in the Sea III: Inputs, Fates, and Effects (http://books.nap.edu/catalog/10388.html)

51. http://seeps.geol.ucsb.edu/

52. *Oil in the Sea III: Inputs, Fates, and Effects* (http://books.nap.edu/catalog/10388.html)

53. http://www.oilspillresponse.com/pdf/Educational/amoco.pdf

54. *Chill - A Reassessment of Global Warming Theory*, Peter Taylor. (Clairview Books, 2009)

55. *The Hockey Stick Illusion: Climategate and the Corruption of Science*, A.W. Montford (Stacey International 2010)

56. International Energy Agency (www.iea.org)

57. *Cool It: The Skeptical Environmentalist's Guide to Global Warming*, Bjorn Lomborg (Marshall Canvendish Ltd. and Cyan Communications Ltd., 2007)

58. International Energy Agency (www.iea.org)

59. *The Hockey Stick Illusion: Climategate and the Corruption of Science*, A.W. Montford (Stacey International 2010)

60. http://joannenova.com.au/2009/07/massive-climate-funding-exposed/

61. US Energy Information Administration
    (http://www.eia.doe.gov/iea/overview.html)

62. *IPCC special report on Carbon Dioxide Capture and Storage*, prepared by working group III of the Intergovernmental Panel on Climate Change

63. Michael Economides and Christine Ehlig-Econmodies, "Sequestering Carbon Dioxide in a Closed Underground Volume", Paper SPE 124430 presented at the Annual Technical Conference and Exhibition of the Society of Petroleum Engineers, New Orleans, 4-7 October, 2009)

64. Lake Nyos disaster, Cameroon, 1986: the medical effects of large scale emission of carbon dioxide?, *BMJ*, Volume 298, 27 May 1989

65. http://www.epa.gov/clearskies/captrade.html

66. "CARBON DIOXIDE CAPTURE FROM FLUE GAS USING DRY REGENERABLE SORBENTS", QUARTERLY TECHNICAL PROGRESS REPORT, DOE Cooperative Agreement No. DE-FC26-00NT40923

67. http://en.wikipedia.org/wiki/Photosynthesis

68. http://www.washingtontimes.com/news/2009/mar/08/carbon-offsets-scam/

69. National Academies, "Understanding and Responding to Climate Change," 2008

70. Michael Moyer, "Fusion's False Dawn," *Scientific American*, March 2010

71. N. Lewis, http://nsl.caltech.edu/energy

72. D. Biello, "R.I.P. hydrogen economy? Obama cuts hydrogen car funding," *Scientific American*, May 8, 2009

73. U.S. Energy Information Administration / Annual Energy Outlook 2010 (www.eia.doe.gov)

INDEX

A

Abiogenic Theory   21-2
AC *(alternating current)*   164-5, 184-6
Accidents   123, 178, 221, 233
Acid   2, 84-6, 88
Acid fracturing   88
Acid rain   147, 152-3, 175, 251
Acid stimulation   83-5
AEF *(America's Energy Future)*   261-6, 270
Africa   23, 25, 34-5, 37, 102-3, 129, 141, 149, 172
Agbami   76-7
Agriculture   176-7, 207
AGW *(anthropogenic global warming)*   15, 220, 237-8, 242, 247, 261, 276
Air   18, 134, 153, 157, 166, 193, 199-200, 206, 209, 229, 244, 248, 251, 266
    compressed   171
Air pollution   125, 230
Aircraft   194-5, 244
Alcohol   85, 194, 197-8, 201, 203, 273
Algeria   32, 36, 38, 98, 108, 112, 155
Alternating current, *see AC*
Alternative fuels   60, 191-3, 195, 198, 207, 215-16
Alternatives   12, 58, 60, 193-4, 244, 260
America   11, 23, 25, 35, 37, 51-2, 116, 134, 150, 269
America's Energy Future, *see AEF*
Amoco Cadiz   105-6, 225, 231
Anglo-Persian Oil Company   29-30
Anthracite   20, 24
Anthropogenic Global Warming, *see AGW*
Antrim shale   41-2
API   4, 19
Ash   93-4, 146-7, 152, 175
Asia   101, 129, 149-50, 173, 175, 178
Asia Pacific   23, 25, 35, 37, 141, 149, 163
Asphalt   22, 115, 120-1
Asphaltene   83, 85

Atmosphere   65, 110, 146-7, 192, 228, 238-9, 242, 244-5, 251-4, 264
Atmospheric CO_2 levels   238-40, 242-3
Australia   24, 44, 55, 148, 150-1, 154, 97, 271

B

Bacteria   198-9
Bakken shale   42-3
Banks   250, 253, 255
Barnett shale   41-2
Batteries   58, 147, 170, 194, 213-15
Battery pack   211-12
Bauxite   87
Biobutanol   198-9
Biodiesel   131, 200-4, 206, 214, 261, 275
Biofuels   3, 59-60, 131, 146, 151, 174, 194-5, 197-9, 203-4, 261, 264
Biogenic Theory   20-1
Biomass   146, 162, 207, 273-4
BioMass-to-Liquids, *see BMTL*
Birds   11, 220, 226
Bituminous coal   20, 274
Blackouts   111, 168-9
Blends   195, 197, 275
Blowout   224, 227, 229
Blowout preventer, *see BOP*
BMTL *(BioMass-to-Liquids)*   207
BOP *(blowout preventer)*   221-2
Borehole   68
BP   4, 11, 29, 31, 39-40, 80, 124, 126, 199, 221-3
BP spill   224, 229
Braer   105-6, 226
Brazil   33, 37-8, 49, 125, 195-7, 204
British Petroleum   29-30
Bunsen, Robert   134-5
Business   45, 55, 57, 75, 131, 168, 175-6, 180, 237, 250-1
Butanes   17-18, 94, 115, 137, 140, 208
Butanol   198-9, 214
Byproducts   59, 115, 117, 146
    coal combustion   147

C

Canada 19, 32-4, 36, 38, 43, 45, 50-1, 55, 95, 98-9, 112, 155, 197, 230, 249

Cap-and-Trade 250, 252

Capacity 101, 108-9, 123-5, 141, 170, 177-8, 184-5, 213, 260

Carbon 15-16, 18, 20, 56, 147, 152, 208-9, 245-6, 259, 263, 266

Carbon capture and sequestration, *see CCS*

Carbon capture and storage 154-5, 246

Carbon credits 249, 254-5

Carbon dioxide 15, 17-18, 56, 136-7, 151-2, 155, 192, 205, 209, 237-9, 242, 245, 248, 254-5, 264

Carbonate formations 85-6, 88

Cars 52, 55

Carson, Rachel 219-20

Cartel 30, 32

Casing 68-9, 71, 78, 82, 84, 86

CBM *(coal bed methane)* 22, 41, 43, 146, 270

CCS *(carbon capture and sequestration)* 153-6, 246-7, 263

Cement 68-70, 232

Cementing 78

CFCs 252-3

Chernobyl 178-9

Chevron 30-1, 39, 77, 80, 126

China 22, 24, 36-8, 44, 49-51, 54-8, 99, 125-6, 147-8, 150-2, 154, 161-2, 173-8, 180, 266, 268

CITGO 130

Cities Service 130

Clean Coal Research 145, 152, 158

Climate 21, 239, 241, 243-4, 254, 261

Climate Change 152, 235, 237-8, 240, 242, 244, 246, 248, 250, 252, 254, 256, 259, 261-62, 266

CNG *(compressed natural gas)* 109-11, 131, 208-9, 268-9, 275

CO$_2$ 15, 17-18, 136, 138, 151, 153-7, 192, 209, 237-9, 241-9, 252-5, 261, 263-4, 266

CO$_2$ emissions 244-5, 247-8, 252, 262-4, 266

Coal 3, 16-17, 20-2, 24-5, 54-5, 58-60, 93-4, 96, 116, 145-54, 156, 158, 162-4, 174-7, 207, 266-70, 273-6

Coal Bed Methane, *see CBM*

Coal Beds 22, 132

Coal Gasification 156

Coal Liquefaction 58

Coal Reserves 24, 148

Coal-to-Liquids, *see CTL*

Coal Tree 146-7

Coconuts 195, 202

Combustion engine, internal 93, 96, 210-13

Combustion reaction 15-16

Companies 29-30, 40, 44-5, 70-1, 78, 80, 101, 124, 126, 133, 176, 206, 223, 254, 274

Compressed Natural Gas, *see CNG*

Condensate 38, 94, 134, 138, 140

Conductors 165-6, 186

Conservation 50-1, 55, 131, 265

Construction 22, 69, 102, 178, 274-5

Consumers 94, 97-8, 100, 111, 161, 164-5, 173, 175, 182, 185, 210, 229, 246, 266

Consumption 50, 54, 150, 171, 173, 191, 203, 242, 249-50

Coolidge, William David 117-18

Corn 194, 198, 203-4, 260-1

Countries 24, 36, 49-50, 53-4, 56-7, 95, 104-5, 148, 151, 161-3, 172-3, 177-8, 180, 185, 188, 206-8

    developed 53, 146, 166, 175, 185

    developing 53, 56, 146, 152, 162

    non-OECD 173-4, 269

    producing 38, 98, 100, 109, 191, 252, 267

Credits 116, 251, 254-5

Crops 176, 204, 243-4, 275

Crude Oil 3, 9, 15-16, 18-21, 33, 41, 74, 94, 100, 102, 107-8, 115, 119, 121-3, 130-1, 223-7

CTL *(coal-to-liquids)* 207, 273

D

DC *(direct current)*   164
DDT   219-20
Deepwater   5, 7, 76-7
Deepwater Horizon   11, 223, 227, 229
Denmark   163, 169, 180
Department of Energy, *see DOE*
Deregulation   132-3
Diesel   16, 52, 96, 115, 120, 123, 191, 193, 200, 203, 208, 260, 269, 273
Direct Current, *see DC*
Disaster   77, 123-4, 233
Discharge   177, 214-15, 219, 221
Disparity   81, 171, 173
Distillation   117, 120, 122, 199-200
Distribution   35, 73, 99, 108, 132, 164, 188
Distribution chain   132-3
Distribution infrastructure   95-6, 175, 180
DMF   199-200, 214
DOE *(Department of Energy)*   53-5, 57, 156-7, 264, 274
Drake Discovery   5
Drilling   5, 43, 66, 70-1, 78, 82, 89, 221, 229, 233, 271
Drought   176-7, 201

E

Earth   22, 66, 71, 182, 184, 186, 220, 226, 237
EIA *(Energy Industry Administration)* 3-4, 9, 33, 39, 49, 51, 53-55, 57, 121-3, 128, 133, 138, 148, 151, 161, 170, 172-4, 267
Electric vehicles   209-11, 213-15
Electrical power   164-6, 185, 213, 215
Electricity   20, 58-60, 100, 111, 146, 156, 165-6, 168-75, 177-8, 183, 185, 209, 249-50, 255-6, 263, 275-6
Electricity consumption   169, 173
Electricity demand   166, 173-4
Electricity generation   59, 146, 156, 162-3, 185, 188, 192, 273
Electrification   164, 172-3, 185
Electrons   165-6

Emissions   147, 151-2, 155-6, 205, 242, 245, 250-3, 264
Energy   15, 49-53, 55-9, 145, 152, 165-6, 170, 181-2, 191, 199-200, 204, 209, 237-9, 259-62, 268-70, 276
   electrical   170, 184, 188, 215
   nuclear   177, 179
Energy business   15, 174, 184, 270, 273
Energy companies   12, 57
Energy conservation   53, 259
Energy consumption   49-51, 54, 192, 265
Energy demand   15, 259
Energy density
Energy efficiency   131, 174, 265
Energy independence   58, 60, 208, 261, 268
Energy industry   10-11, 61, 220
Energy Industry Administration, *see EIA*
Energy Intensity   51-3, 57
Energy interchangeability   58, 60-1
Energy mix   54, 273
Energy requirements   184, 233, 276
Energy resources   55, 58, 132, 273
Energy security   192, 195
Energy sources   58, 65, 145, 147, 181, 259-60, 262, 264
   alternative   4, 55, 60, 141
   primary hydrocarbon   16
   traditional   254
Energy supply   12, 96, 255, 259, 275-6
   primary   145, 191
Energy taxes   249-50
Energy Tribune   8, 59-60, 89
Engine   59, 94, 193, 195, 200
   coal-fired external combustion   93
   compression-ignition   200, 202
Enhanced Oil Recovery, *see EOR*
ENI *(Ente Nazionale Idrocarburi)*   31
Ente Nazionale Idrocarburi, *see ENI*
Environment   70, 75, 119, 152, 197, 217, 219-22, 224, 226-32, 234, 262
Environmental groups   193, 204, 220, 237, 244

Environmental Protection Agency, *see EPA*

EOR *(enhanced oil recovery)* 140, 155, 247

EPA *(Environmental Protection Agency)* 125, 253

Ethane 17-18, 94, 115, 134, 136-7, 140

Ethanol 16, 59-60, 131, 194-200, 202-4, 275

Ethanol production 197, 204

Eurasia 23, 35, 174

Europe 35, 58, 98-9, 129, 131, 141, 154, 163, 174, 179, 188, 210, 266, 271

Expansion 41, 128, 132-3, 141, 152

Exploration 22, 40, 66, 81, 131-2, 220-1, 233, 265

Explosions 11, 215

Explosives 147, 224

Extraction 132, 138, 265, 270

Exxon 30-1, 45

Exxon Valdez 105-7, 221, 223-4, 226, 228, 230-2

F

Facilities 9, 70, 91, 93-4, 96, 98, 100, 102, 104, 106, 108, 110, 112, 171, 174, 176

Factories **96, 196, 233, 244, 254**

Failure 11, 56, 168-9, 215, 264

Faults 66, 71, 228

Fayetteville shale 41-2

Feedstocks 115, 121-2, 140, 206, 274-5

Fermentation 198, 200, 205

Fertilizers 147, 204, 219, 244

Fines 82, 84-6

Fischer-Tropsch Process 207

Fixed platform, *see FP*

Flaring 111-12

Floating Production, Storage And Offloading, *see FPSO*

Flowlines 74, 76, 78

Fluids 21, 41, 78, 82, 85-6, 139

Food prices 204, 260-1

Forecasts 54-5, 180, 203, 265, 267-70

Formation 29, 79, 82-3, 85-8, 110, 139, 157
    geologic 247

Formation damage 82, 84

Fossil Fuels 3-4, 6, 8, 10, 12, 14, 16, 18, 20-2, 24, 54, 58-60, 145, 178, 204, 265-6

FP *(fixed platform)* 75-6

FPSO *(floating production, storage and offloading)* 8, 75-6, 78

Fractions 119, 129, 191, 255, 276

Fractures 22, 85-7, 228, 272

Fracturing 82, 84, 88
    hydraulic 86, 88-9

Fracturing Fluid 85, 87-8

France 31, 49, 51, 98, 126, 155, 169, 180, 194, 249

Fuel oils 115, 193
    distillate 122-3

Fuel sources 132, 151

Fuels 9-10, 15-16, 19, 24, 59-61, 93-5, 119-20, 140, 145-7, 175, 191-5, 199-200, 202, 204-9, 214-15, 249-50
    conventional 193, 204-5
    diesel 117, 121, 123, 131, 153, 200
    primary 146, 163
    renewable 25, 203
    transportable 132

G

Gas Hydrates 139

Gas pipelines 98, 134

Gas prices 45, 53, 163, 191
    natural 81, 148, 269

Gas processing 137, 140-1

Gas production 7-9, 82, 139
    natural 7, 35, 41, 43, 137, 270

Gas reserves 22, 34, 37, 95, 271
    natural 23, 25, 34-5, 39, 132, 270

Gas reservoirs 74, 138, 228, 246-7, 272

Gas shales 88

Gas turbine 111, 156-7

Gas wells 67-8, 82, 139

Gasifiers 273-5

Gasohol 194-5, 197

Gasoline 9, 16-17, 53, 59, 96, 115, 117, 120-1, 123, 193-5, 198-200, 203, 207-10, 213-14, 249, 269
    natural 137, 140
GDP *(gross domestic product)* 49, 51-3, 56, 255
Generating Capacity 111, 170, 175, 178, 184
Generating electricity 146, 158, 163, 175, 182-3
Generation 132, 166, 175, 177, 188, 237
Geothermal sources 185
Germany 44-5, 49, 51, 98, 150, 155, 163, 171, 180, 195, 249, 273
Gesner, Abraham 116
Global temperatures 263
    inferred 239, 241
GoM, *see Gulf of Mexico*
Governments 56-7, 65-6, 177-8, 184, 201, 223, 240-1, 246-7, 249, 253
Greenhouse gas emissions 95, 153, 183, 204, 264
Greenhouse gases 110, 152, 155, 239, 254, 263-4
Grid 168-71, 181, 184, 215, 260
    electrical 163-4
Gross Domestic Product, *see GDP*
Ground 19, 67, 71, 153, 184-5, 207, 247
Growth 29, 40-1, 45, 117, 173, 175, 180-1, 205, 269, 276
GTL 207, 273, 275
Guangxi 176-7
Gulf of Mexico *(GoM)* 8, 11, 40, 77, 221-9, 232-3

H
HCl 84-5, 88
Heat 15-16, 20, 59, 166, 184, 206, 239
Heavy crudes 123, 130
Heavy oil 19, 21, 45, 130
Helium 18, 136-7, 140
HFC-23 254
Hg 17, 136, 139

Homes 59, 134, 166, 168, 176, 215
HVDC 111, 186-8
Hybrid vehicles 211-14
Hydrates 110-11, 139
Hydrocarbon components 17-18, 119
Hydrocarbon fuels 15, 17-18, 100, 200, 244
    conventional 215
    major 162, 200
    solid 20
    traditional 204
Hydrocarbon gases 136, 208
Hydrocarbon resources 15, 276
Hydrocarbons 3, 15-18, 20-2, 25, 41, 65, 67, 72-3, 77-8, 82, 85, 207, 219-20, 226-9, 231, 275-6
Hydrochloric acid 84-5, 88
Hydrocracking 120, 207
Hydroelectric power plants 177
Hydroelectricity 176-7
Hydrofluoric acid 84-6
Hydrogen 15-17, 20, 58-9, 120, 134, 153, 156, 194, 207, 209, 214, 273
Hydrogen sulfide 17-18, 136-7, 248
Hydrotreating 120

I
Ice Ages 239, 243
ICE *(internal combustion engine)* 93, 210
IEA 161-2, 202
IGCC *(Integrated Gasification Combined Cycle)* 153, 156
Importing Countries, Major 58, 108
Impurities 17, 20-1, 136-7, 183
Incidents 178, 221-2, 227, 232-3
    tanker 105, 107
Independents 44-5
India 24, 30, 44, 49, 125-7, 148, 150-2, 154, 161, 173, 175, 178, 180, 197, 266
Indonesia 32, 36, 38, 49, 51, 108, 112, 126, 154, 161, 185
Industry 4, 7, 10-11, 14-15, 50, 59, 67, 78, 81, 96, 119, 124-5, 128, 223, 241-2, 274-5

auto 117
aviation 194
refinery 116-17, 124
Infrastructure 58, 94, 141, 175, 208, 214
Injection 19, 86, 153, 156, 247, 272
Installed capacity 163, 180, 185
Integrated Gasification Combined Cycle, *see IGCC*
Integrated Oil Companies, *see IOCs*
Internal Combustion Engine, *see ICE*
Invention 117, 134, 244
Investment 3, 44, 55, 95-6, 129-30, 162, 169, 174-5, 184, 214
IOCs *(integrated oil companies)* 14, 29
IPCC 238, 241, 251
Iran 22-3, 32, 34, 36-8, 55, 95, 112
Iraq 23, 32, 34, 36, 38, 112
Iso-)ctane 16-17
Italy 30, 49, 51, 98, 169, 180, 249

J
Jackup 67
Japan 51-3, 95, 108, 110, 126, 131, 150, 154-5, 177, 249
Jatropha 201, 203-4
Jet fuels 115, 191, 273
Jevons Paradox 50-1

K
Kalfayan 83-4
Kazakhstan 24, 36, 99, 112, 151
Kerosene 16, 100, 115-17, 122, 193
Kuwait 23, 32, 34, 36, 38, 112, 221, 224

L
Lake Nyos in Cameroon 248
Land 7, 66-7, 70, 74, 111, 176, 186, 193, 204
Leakage 65, 178, 223, 229, 246, 248
Leaks 223, 227, 232-3
Libya 32, 34, 38, 98, 112
Life 12, 15, 57, 78, 124, 200, 219, 237-9, 243-4
Light 19, 21, 41, 122-3, 134, 165, 177, 182-3, 205, 214

Lignite 20, 24, 274
Liquefied Natural Gas, *see LNG*
Liquid fuels 60, 93, 205, 207, 265, 275
Liquid hydrocarbons 134, 136
Liquid Petroleum Gas, *see LPG*
Liquid transportation fuels 59-60, 207
Liquids 18, 54-5, 59, 74, 93-4, 108, 121, 136, 151, 184, 202, 208, 215, 267-70, 273
LNG *(liquefied natural gas)* 95-6, 99, 108-9, 111, 175
LNG tanker 108-9, 132
LPG *(liquid petroleum gas)* 115, 120, 131, 208
Lukasiewicz, Ignacy 117

M
Macondo 77, 232
Maize 194, 198
Malaccamax 104
Malaysia 13, 36-8, 108, 112
Mancos 42
Manufacture 140, 199-200, 219-20, 254
Manufacturing 10, 14
Marcellus shale 42-3, 271
Markets 54, 67, 108, 110, 129, 154, 191, 194, 202, 250-1, 255
Mattei, Enrico 30-1
Media 1, 58, 177, 193, 226, 232-3, 240, 259, 276
Mercury 17, 136, 139, 156, 175
Mergers 29-32, 41
Methane 17-18, 22, 41, 43, 94, 108, 110-11, 115, 134, 136-7, 140, 146, 153, 207-8, 246, 270
Methanol 16, 139-40, 199, 201, 273
Mexico 33, 38, 49, 99, 112, 126, 185
Michigan 85, 168-9
Middle East 23, 25, 31, 34-7, 45, 129, 141, 148-9, 162, 172
Mobil 30-1
Money 208, 254-6, 276
Motor, electric 211-13
Multilateral wells 70, 72
Myths 39-40, 260

N

Naphtha   100, 115, 117, 122, 207

National Iranian Oil Co.   39-40, 126

National Oil Companies, *see NOCs*

Nations   50-2, 55, 58, 261, 265-6

Natural Gas   3, 16-18, 20-2, 24, 34, 54-5, 59-60, 94-5, 98-9, 108-10, 132-6, 139-41, 148-9, 162-4, 207-9, 267-70

    adsorbed   146

    compressed   109, 131, 208

    conventional   111

    green   276

    pipeline quality   136

    recoverable   22

    synthetic   153

    unconventional   41, 44, 266

Natural Gas Hydrates, *see NGH*

Natural Gas Liquids, *see NGLs*

Natural gas processing   9, 132-3, 135-6, 139, 141

Natural gas seeps   22, 228

Natural gas storage   155

Natural gas utilization   176

Natural gas vehicles   208

Netherlands   38, 44, 50, 127

New technologies   21, 24, 95, 156, 250-1, 264, 274

NGH *(natural gas hydrates)*   110-11, 266, 268

NGLs *(natural gas liquids)*   18, 38, 133, 136-8, 140

NGWDA *(Natural Gas Wellhead Decontrol Act)*   132

Niagara   165

Nigeria   32-3, 36, 38, 76, 108, 112

Nitrogen   15, 17-21, 136, 138, 175

NOAA   232

NOCs *(national oil companies)*   29, 37, 40, 45, 66, 81

Non-OPEC   33-4

North America   19, 34-5, 41, 43, 98, 129, 141, 149-50, 163, 271

Norway   36, 38, 98-9, 155, 188

Nuclear   54-5, 60, 146, 151, 162, 192, 267-70

Nuclear power   163, 170, 177-8

O

Oceans   21, 65, 109, 227-9, 239-40, 245

OECD   161, 162, 171-3, 268

OEX   80

Offsets   **67, 249, 251**

Offshore developments   44, 230

Offshore drilling   11, 222-3, 229

Offshore fields   74, 77

Offshore oil production   7, 230

Ohio   85, 168-9

Oil   16-22, 24-5, 29-30, 36-45, 58-60, 65-9, 71-6, 81-4, 93-8, 134-8, 141, 148-51, 191-3, 203-6, 219-22, 228-33

Oil companies   65, 80-1, 100, 246-7

    international   40

    major   30, 45

    national   40, 66, 81, 130

Oil exploration   227, 229

Oil pipelines   97, 99

Oil prices   19, 30-2, 67, 148, 191, 196-7, 206, 208, 229-30, 269

Oil production   4, 8, 32, 117, 131

Oil Refining and Gas Processing   113, 115-16, 118, 120, 122, 124, 126, 128, 130, 132, 134, 136, 138, 140, 142

Oil reserves   23-4, 40, 58, 210

Oil seeps   227-8

Oil spills   107, 223, 227

Oil storage tanks   104-5

Oil tankers   74-5, 100, 102, 104, 108

Oil wells   82, 85, 224

OPEC *(Organization of the Petroleum Exporting Countries)*   31-4

organisms   226

Organization of the Petroleum Exporting Countries, *see OPEC*

OSX   80

Otters   226

Output   121-3, 125, 156, 163, 184

Oxygen   15, 17-18, 20, 152-3, 156, 202, 209, 252

Ozone   252

P

Paraffin   83, 85

PdVSA *(Petróleos de Venezuela, S.A.)* 37, 126, 130

Pennsylvania   5

Pentane   18, 136-7, 140

Permeability   41, 73, 88

Petrodiesel   200-2

Petróleos de Venezuela, S.A., *see PdVSA*

Petroleum   18, 120, 192, 216, 228, 260-1, 264

Petroleum coke   115, 121-2, 273

Petroleum products   12, 20, 29, 33, 115

Petroleum tankers, *see oil tankers*

Petronas   37, 39

Photovoltaic   166, 182-3

Pipelines   74, 76-7, 93-8, 100, 108, 111, 132-3, 135, 137-40, 194, 214, 266

Pipes   68, 139

Planet   110, 181, 207, 237-9, 243-5, 253

Plants   133, 147, 165, 170, 175, 178, 183, 185, 192, 204, 206, 253, 263-4

    underground coal gasification   154

Plastics   14, 59-60, 147, 219, 244

Platform   8, 70, 74-6

Poland   44-5, 150, 271

Politicians   11, 177, 232, 237-42, 244, 276

Pollutants   18, 194, 250, 253

Pollution   12, 56, 104, 125, 175, 213, 217, 219-20, 222, 224, 226, 228, 230, 232, 234, 250-1

Population   172, 175, 219

Ports   102, 104

Portugal   180

Power   10, 30-2, 58, 111, 131-2, 152, 156, 163, 166, 168-9, 175-7, 181-2, 186, 188, 244-5, 275-6

Power generation   50, 60, 147, 153, 158-9, 161-2, 164-6, 168, 170, 172, 174-7, 180, 182, 184, 186, 269

    electrical   146, 163, 183

    nuclear   177-8

Power outages   168-9

Power plants   155, 170, 175, 177, 205

Power stations   166, 169, 246, 253

Prestige   105-6

Price   59, 67, 108, 148, 186, 191-2, 210, 255

Producers   45, 98, 100, 111, 197

Production   7, 25, 37-8, 65-6, 70, 76, 78, 80-2, 88-90, 131-2, 140, 149-50, 197-9, 202-3, 229-30, 252-4

Production capacity   43, 207

Production facilities   7, 76

Products   14, 108, 122, 134, 136-7, 141, 157, 175, 200-1, 250, 253

Projects   41, 70, 131, 154-6, 165, 188, 242

Propane   17-18, 94, 115, 134, 136-7, 140, 208

Proppant   84, 87-8

PV cells   183-4

Pylons   185-6

Q

Qatar   5, 22-3, 32, 36, 38, 95, 108, 112

R

R/P *(reserves to production ratio)*   36-7

Raw materials   10, 140-1, 199, 219-20, 233, 275

Refined products   74, 93, 100, 104, 117, 121-3, 129

Refineries   74, 117, 119, 121-5, 127-8, 130-1, 226

    conventional   206

    largest   125, 127

    new   117, 124-5, 141

Refining   9, 18, 29, 115, 117, 123-4, 130-1, 133, 135-6, 141, 220, 249

Regions   24, 36, 129, 131, 166, 169, 173

Renewable sources   25, 176, 200, 205, 207

Renewables   54-5, 60, 151, 265, 276

Research   240, 255-6, 261, 269

Reserves   22-4, 36-7, 40, 45, 76, 148-9
Reserves to Production Ratio, *see R/P*
Reservoirs   5, 21, 41, 70, 177, 215, 248
Resistance   50, 166, 185
Resources   3, 12, 22, 45, 50, 219, 266
    hydrocarbon fuel   22
    natural hydrocarbon   233
    stranded   96, 111
    unconventional   41, 43, 264
Rigs   67, 221
Risk   77, 105, 110, 123, 178, 215, 220, 233, 243, 248, 254
Road oil   121-2
Rock formation   85-6
Rocks   41, 66, 72, 82, 105, 117, 232
Royal Dutch Petroleum Company   29-30
Royal Dutch Shell   30-1, 40, 102, 126
Russia   22-4, 32, 34, 36-8, 45, 49, 95, 98-9, 112, 125-6, 148, 151, 154, 271

S

Salt deposition   82-3
Samuel, Marcus   97, 100-2
Sand   5, 78, 87-8, 139
Sandstones   21, 41, 82, 85-6
Santa Barbara   228
Saudi Arabia   22-3, 32-4, 36-8, 112
Saudi Aramco   37, 40, 126
Scale deposition   82-3
Schlumberger   42, 80-1, 85
Scientists   15, 193, 231, 233, 239-42, 261
Scrubbers   139
Sea   11, 70, 104, 193, 222, 225, 228-9, 241
Sea levels   71, 262-4
Seaborne Oil Trade   107
Seawaymax   104
Seawise Giant   102, 104
Sectors   61, 80, 145, 180, 192
Seeps   22, 65, 116-17, 226-9
    natural   227-8, 232
Service companies   66, 72, 81, 222
Services   70, 78, 80-1, 213, 255
    oilfield   81

seismic   78, 80
Seven Sisters   30-1, 37, 39, 45
Shale gas   41, 43-4, 89, 270-1
Shales   5, 44, 88, 176
Shallow-Water, Ultra-Deep, *see SWUD*   43
Shelf, continental   230
Shell   39, 68
Shell Transport and Trading Company   29-30, 102

Ships   52, 93, 96, 100-2, 105, 108-9, 193, 241
    drill   67
    largest   102-3
Siemens, Sir William   153-4
Society   25, 50, 145, 161, 233, 254
Soils   204, 227, 245
Solar   59-60, 162-3, 166, 179, 181, 184, 209
Solar collectors   182, 270
Solar energy   181-2, 263
Solids   111, 115, 135, 138-9
Sour gas   137
South Africa   24, 148, 150-1, 154, 183, 207, 273
South Korea   49, 51, 127, 150
Soviet Union   31-2, 141, 178
SO_x 147, 251-3
Soybeans   194, 203, 261
Spain   49, 51, 98-9, 108, 163, 180
Spill   105, 225-6, 231-2
SS *(sub sea system)*   75
Standard Oil   29-30, 85, 101
Statoil-Hydro   155
Steam   19, 156, 166, 183-4, 207
Stimulation   81-2, 84-5, 88-9
Storage   17, 75, 132, 154-6, 170, 199, 246, 261, 263, 265
Sub Sea System, *see SS*
subsidies   184, 194, 198, 201, 259
Suez Canal   100-1, 103-4
Sugar cane   195-6, 203, 204
Sulfur   15, 19-21, 122, 137-8, 140, 202,

251-3

Sulfur dioxide 18, 56, 152, 156, 251-3

Supertankers 102-3, 231

Surface 19, 21-2, 65-6, 70, 72, 74, 76, 78-9, 97, 139, 184, 221, 226, 228, 232

SWUD (*shallow-water, ultra-deep*) 43

Syngas 153, 156-7, 207, 273, 275

Synthesis gas 273

T

Tankers 100, 102, 108, 225, 231

Tar sands 5, 19

Tars 115-16

Taxes 208, 242, 249, 253

Teamsters 96-7

Technologies 45, 58, 65, 70, 78, 80, 88-9, 95-6, 110-11, 152, 154-5, 163, 205-6, 259, 262-5, 270-4

Temperatures 5, 18, 21, 71, 82, 111, 119, 139, 153, 207, 239-40, 243, 262, 274

Tension Leg Platform, *see TLP*

Tesla, Nikola 164-5, 186

TLP (*tension leg platform*) 8, 75

Torrey Canyon 105-6, 225

Toyota Prius 211-12

Trade 100, 251-3

Traditional Clean Coal Technology 152

Truly Clean Coal Technology 152

Trains 93, 95

Transformers 161, 164, 185

Transmission 66, 168, 175, 180, 185-6, 188, 263

Transport 29, 77, 93-4, 110, 132, 175, 193-4, 242, 244, 249

Transportation 58-60, 95, 145, 192-3, 208, 220, 244, 264-5, 273

Transportation fuel 191, 195, 203, 208

Transportation sector 191-2, 268

Transporting 17, 77, 97, 100-1

Tubing 69, 82-3

Turbines 166, 171, 202, 260

Turkmenistan 36, 38, 99

Typhoon 74

U

UAE 23, 38

UCG (*Underground Coal Gasification*) 153-4

Ukraine 24, 98, 151, 177-8

ULCC (*Ultra Large Crude Carrier*) 102, 104

Ultra Large Crude Carriers, *see ULCC*

Underground Coal Gasification, *see UCG*

United Arab Emirates 22, 32, 36

United States 24, 36, 49-58, 99, 117, 124-5, 133-4, 181, 185, 194, 227, 230, 260-1, 266, 268

US 32-4, 36, 41, 56-7, 123-5, 128, 130-3, 146, 148, 150-2, 169-70, 180, 197, 207-8, 229-30, 270

US Department of Energy 156, 274

USA 53, 98, 108, 112, 127, 171, 175, 177-8, 180, 197

Utsira Formation 155

V

Valero Energy 80, 126

Van Syckel 97

Vegetable oils 201-4, 206

Vehicles 52, 58-9, 93, 95-6, 203, 208-9, 211, 213-15, 275

Venezuela 19, 23, 32-4, 36-9, 45, 112, 126-7, 130-1

Very Large Crude Carrier, *see VLCC*

Vessels 8, 76, 101-2, 104-5, 107, 109

Viscosity 19

VLCC (*Very Large Crude Carrier*) 102, 104

Voltage 166, 185

W

Waste 3, 136, 162, 202, 206, 219

Water 7, 19, 21, 41, 73-4, 78, 82-3, 85, 88, 110-11, 153, 171, 176-7, 199-200, 206, 209

Water depth 7, 75, 221

Wax 82-3, 85

Wellbore 41, 78, 83, 85-8

Wellhead 74, 76, 96, 134, 139, 224

Wells 5, 21, 67-8, 70-1, 78, 81-2, 86, 88, 137, 139, 153, 184, 233, 248, 272

Westinghouse, George 164-5

Wheat   203

Wind   3, 59-60, 146, 162-3, 166, 179-81, 184, 226, 251
Wind Power   163, 180-1
Wind Turbines   180-1, 188, 260, 270
Windmills   179
Wireline   72, 78
Woodford shale   41-2
World Economy   268-9, 276
World Electricity Supply   162
World Energy Demand   3-4, 267-9
World Oil Production   19, 40, 221
World War II   30, 52, 134, 154
WTI   148

X
XTO   43, 45

Y
Yeast strains   198-9
Young, James "Paraffin"   116

Z
ZEEP   273
Zoroaster   100
Zooplankton   21